SpringerBriefs in Molecular Science

Green Chemistry for Sustainability

Series Editor
Sanjay K. Sharma

For further volumes:
http://www.springer.com/series/10045

Dickcha Beekaroo · Ackmez Mudhoo

Adsorption of Reactive Red 158 Dye by Chemically Treated *Cocos nucifera* L. Shell Powder

Adsorption of Reactive Red 158 by *Cocos nucifera* L.

 Springer

Dickcha Beekaroo
Faculty of Engineering
Department of Chemical and
  Environmental Engineering
University of Mauritius
Reduit
Mauritius
e-mail: anyatra19@yahoo.com

Ackmez Mudhoo
Faculty of Engineering
Department of Chemical and
  Environmental Engineering
University of Mauritius
Reduit
Mauritius
e-mail: ackmezchem@yahoo.co.uk

ISSN 2191-5407
ISBN 978-94-007-1985-9
DOI 10.1007/978-94-007-1986-6
Springer Dordrecht Heidelberg London New York

e-ISSN 2191-5415
e-ISBN 978-94-007-1986-6

Cover design: eStudio Calamar, Berlin/Figueres

Printed on acid-free paper

Springer is part of Springer Science+Business Media (www.springer.com)

Dedicated to Neelam

Contents

Adsorption of Reactive Red 158 Dye by Chemically Treated *Cocos nucifera* L. Shell Powder

Abstract The effective removal of dyes from aqueous wastes is an important issue for many industrialised countries. The traditional treatment methods used to remove dyes from wastewaters have certain disadvantages such as incomplete dye removal, high reagent and energy requirements, generation of toxic sludge or other waste products that require further disposal. The search for alternative and innovate treatment techniques has focused attention on the use of biological materials for dye removal and recovery technologies. In this respect, adsorption has gained an important credibility during recent years because of its good performance and low cost performance as a pollutant removal technology. In this research study, chemically conditioned *Cocos nucifera* L. shell powder was used as low-cost, readily available and renewable adsorbent for the removal of reactive textile Red-158 dye from aqueous solutions. Batch experiments were carried out for the adsorption kinetics and isotherms with pre-treated *C. nucifera* L. Operating variables studied were pH, initial dye concentration and dosage of adsorbent. Results indicated that the adsorption capacity had been enhanced with increasing dosage of the biosorbent and the maximum colour removal was 57.18% at a pH of 2.00 ± 0.01 with an adsorbent dosage of 20 g/L and dye concentration of 20 mg/L. Equilibrium adsorption data were very well represented by the Langmuir, Freundlich, Dubinin–Radushkevich and Temkin isotherm models with all having correlation coefficients higher than 0.85 for all tested adsorption systems. R_L values obtained for the Langmuir model were greater than 1 and the n values of Freundlich being in the range of 2–10 collectively indicated that the biosorption process with *C. nucifera* L. was feasible. Equilibrium kinetic data followed the second-order equation and intra diffusion model very well, supported by the high correlations coefficients of the respective linearised models. Batch column experiments were also carried out with operating variables such as bed height, initial dye concentration and input volume of dye solution. The dye uptake was higher with increasing bed heights such that dye removal uptake was 73.60% at a bed height of 6 cm and 92.49% at bed height of 10 cm. At constant bed

D. Beekaroo and A. Mudhoo, *Adsorption of Reactive Red 158 Dye by Chemically Treated Cocos nucifera L. Shell Powder*, SpringerBriefs in Green Chemistry for Sustainability, DOI: 10.1007/978-94-007-1986-6_1, © The Author(s) 2011

1

heights and with constant dye concentrations, the dye removal efficiency decreased as input volume of dye solution to the column rises and so dye uptake in first 120 min was 92.63% for an input of 100 mL dye solution and 60% with higher input volume of 250 mL while an increase in initial dye concentration leads to a decrease in the uptake of reactive red 158 dye. Experimental data exhibited a characteristic "S" shape and could be fitted effectively to the Thomas model which had high correlation coefficients (above 0.85).

Keywords *Cocos nucifera* L. · Reactive Red 158 dye · Adsorption · Thomas model · Langmuir model

1 Introduction

Over the past decades, the exponential population and social civilization expansions, changes in lifestyles and resources used coupled with continuing progress of the industrial and technologies have led to a sharp modernization and metropolitan growth [6]. With the rising awareness of the occurrences of industrial activities, numerous deteriorations on several ecosystems have been intensified and have started to seriously threaten the human health and environment. Among the many instances of pollution of aqueous media, dye contamination of waters occurs when a wide spectrum of chemicals and dyes are discharged directly or indirectly into water bodies without adequate treatment to remove and degrade these harmful compounds. More than 100,000 commercially available dyes are known and approximately 1 million tons of these dyes are produced annually worldwide. The major sources of dye contamination are considered to be from textiles industry [7].

Wastewater from the textile industry is a complex mixture of many polluting substances ranging from residual dyestuffs to heavy metals associated with the dyeing and printing processes. Wastewater that originates from reactive dye processes is particularly problematic because such dyes have low levels of fixation to the fibers. 30% or more of the dyes used are hydrolyzed and then released into waterways. The brightly coloured unfixed dyes are highly water soluble, have poor adsorption properties and are not removed by conventional treatment systems. Although these dyes are not in themselves toxic, they may be converted into potentially carcinogenic amines after release into the aquatic environment. Their disposal is always a matter of great concern since they are considered as a quite dangerous source of environmental pollution and may cause direct destruction of aquatic life due to the presence of aromatic and metal chlorides which interferes with light penetration and oxygen in water bodies leading to conflict between upstream discharger and downstream user water. The dyes comprise likewise an aesthetic problem. In order to comply with the environmental standard limits, it is essential that industries treat their effluents to reduce the dyes level to acceptable levels, and in the same time also prevent further irreversible damage to the environment at large.

A number of treatment methods for the removal of dyes from textile effluents that are most commonly used are mainly reduction, ion exchange, electro-dialysis, electrochemical precipitation, evaporation, and chemical precipitation. However, these techniques are non-destructive, since they only transfer the non-biodegradable matter (dyestuffs) into sludge, thus giving rise to a new type of pollution, which needs further treatment. In principle, it is very difficult to treat dye laden waste-waters due to their high solubility and complex molecular structures. Each treatment process has its own limitations such that they suffer from drawbacks such as high capital and operational costs or the disposal of the residual metal sludge. However, the adsorption technique is being considered to be a most appropriate technique. This is because adsorption is relatively simpler, much selective, cost-effective, and easy to operate and a proven efficient process for the removal of dyes from contaminated aqueous media. Previous work has been carried out on the use of several adsorbents such as activated carbon, low cost agricultural waste materials and microbial biomass for the removal of many kinds of dyes. Activated carbons with their large surface area, micro porous character and chemical nature of their surface area and high adsorption capacity are potential adsorbents for the removal of dyes from industrial wastewaters. But the high costs make the process uneconomical for industrial applications and hence, the process of dye removal by adsorption is being diverted to the use of low-cost adsorbents so that the process becomes economically feasible [4]. For this reason, research is now focusing on the use of low-cost, reusable, locally available, biodegradable adsorbent made from natural sources like rice husk, groundnut husk carbon, coconut husk and palm pressed fibres, coconut shell activated carbon (CAC's) and many others.

Coconut shell of the coconut species *Cocos nucifera* L. is a material composed of several constituents consisting of lignin and hemicellulose. These bear various polar functional groups including carboxylic and phenolic acid groups which can be involved in metal and dye molecules binding. The cellulose and lignin are biopolymers admittedly associated to the removal of dyes and thus, *C. nucifera* L. has been tested in this research work. The main aim of this study has been to probe into the adsorption characteristics of Reactive Red 158 dye (RR158) from synthetic aqueous solutions using chemically activated *C. nucifera* L. shell powder sorbent. The specific objectives of the study were as follows:

- Preparation of chemically activated *C. nucifera* L. shell powder, hereinafter referred to as the "adsorbent"
- Determination of the yield of adsorbent obtained from the chemical treatment of the native coconut shell
- Determination of the optimum wavelength of RR158 solutions at which its absorbance is maximum and the calibration curve of absorbance and concentration of dye solution
- Determination of the optimum pH at which the batch adsorption experiments involving RR158 and the adsorbent have to be carried out

- Studying the effects of dosage of adsorbent on the removal of RR158 and also quantify the removal efficiency for various dosages
- Studying the effects of initial RR158 concentrations on the removal of RR158 and also quantify the removal efficiency for various concentrations
- Analysing the pool of results obtained using the standard equilibrium isotherms namely Langmuir, Freundlich, Temkin and Dubinin–Radushkevich models
- Studying the kinetics of the adsorption behaviour of RR158 for the said adsorbent using the pseudo first order, pseudo second order equation and Intra particle diffusion models
- Studying the adsorption behavior of red dye 158 with the CAC's in batch column studies
- Studying the effects of initial RR158 concentrations on the removal of RR158 and also quantify the removal efficiency for various dosages in batch column studies
- Studying the effects of bed height and input volume on the removal of RR158
- Analysing the results obtained for the column studies using Thomas model
- Data analysis using error functions of residual sum of squares, average relative error and sum of absolute errors for both the equilibrium isotherms and kinetics equations

2 Literature Data

2.1 Synthetic Dyes

A dye is a coloured compound that can be applied on a substrate by one of the various processes of dyeing, printing, surface coating, and so on. Most of the dyes have been made from natural sources such as various parts of plants or certain animals until synthetic dyes were developed in the late 1800s and early 1900s. Most of the synthetic dyes are aromatic organic compounds and generally, the substrate that is the material on which dye is being applied includes textile fibers, polymers, foodstuffs, oils, leather, and many other similar materials. Dyestuffs are produced over 700,000 tons annually estimated from more than 100,000 commercially available dyes and applied in many different industries. A dye molecule consists of two major elements known as chromophores and auxochromes whose structures. Chromophores (*"chroma"* means colour and *"phore"* means bearer) is an unsaturated group basic that is responsible for colour while auxochromes (*"Auxo"* means augment) are the characteristic groups which intensify color or improve the dye affinity to substrate. Dyes are classified firstly, according to their chemical structure particularly considering the chromo-phoric structure presents in the dye molecule and secondly based on their usage or application (Fig. 1).

RR158 also known as Levafix brilliant red E4B4 ($C_{29}H_{19}O_{11}N_7S_3ClNa_3$) utilises a chromophore containing substituent that is capable of directly reacting

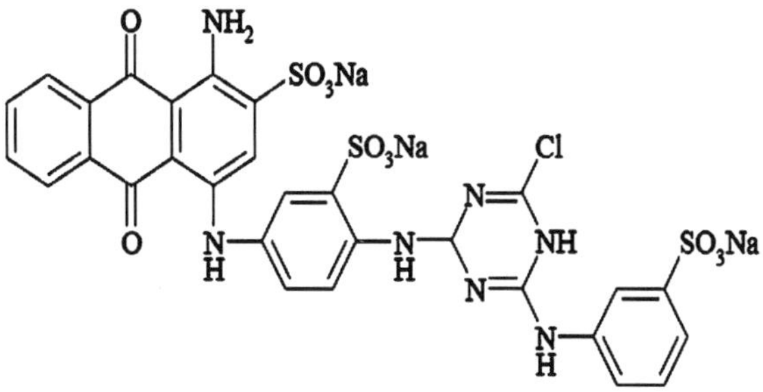

Fig. 1 Structure of Reactive
Red-158 dye

Table 1 Characteristics of
RR-158 dye

Parameter	Value
pH	6.3
Wavelength (nm)	513
Molar absorptivity (mol/L cm)	1.4×10^4
Limit quantification (blank, mg/L)	0.35
Beer's law range (mg/L)	0.5–15.0
Slope	0.0141
Intercept	−0.0195
R^2	0.999

with the fiber substrate. The covalent bonds that attach reactive dye to natural fibers make it among the most permanent of dyes. Cold reactive dyes, such as Procion MX, Cibacron F, Levafix Brilliant and Drimarene K, are very easy to use because the dye can be applied at room temperature. Due to their strong interaction with many surfaces of synthetic and natural fabrics, reactive dyes are highly used for dyeing of wool, cotton, nylon, silk, and modified acrylics. Some classes of dyes are harmful to aquatic life even at lower concentrations. It is pointed out that less than 1.0 mg/L of dye content causes obvious water coloration. Dye concentrations of 10 mg/L up to 25 mg/L have been cited as being present in dyehouse effluents. After mixing with other water streams, the concentration of dyes is further diluted. The limit of concentration of some toxic dyes in water is 1.0 ng/L. Table 1 provides information about the characteristics of RR158 dye.

2.1.1 Dye as a Water Pollutant

The discharge of highly coloured synthetic dye effluents is an indicator of water pollution and it is aesthetically displeasing. The World Bank estimates that 17–20% of industrial water pollution comes from textile dyeing and treatment. They have also identified 72 toxic chemicals in our waters solely from textile dyeing, 30 of which cannot be removed (Khehra et al. 2005). Dye wastewaters contain many toxic organic residues. Their disposal is always a matter of great concern since they are considered as a quite dangerous source of environmental pollution that may cause direct destruction to aquatic life due to the presence of aromatic and metal chlorides. Moreover, colorant may interfere with light penetration and oxygen in water bodies decreasing the biological oxygen demand (BOD) level in the water streams. The dyes comprise likewise an aesthetic problem and colour restricts which lead to conflict between upstream discharger and downstream user water. Thus, effective and economical treatment of effluents with diversify textile dyes has become a necessity for clean production technology for all industries.

Water pollution by dye contamination may arise from different manufacturing industries such as textiles, paper, cosmetic, leather, food, paint, electroplating, galvanized and powder batteries processing units and pharmaceutical industries. The textile industry is considered to be the largest group using synthetic dyes constituting of 60–70% of all dyestuffs [10]. Textile dyes form a large group of textile chemicals and comprise over 8,000 different compounds with almost 40,000 commercial names. The textile industry is the largest consumer of dyes with a consumption of around 60% of that produced using them in conjunction with a wide range of auxiliary reagents for various dyeing, printing and finishing processes. About 20–50% of the reactive dye can be released into waterways depending upon dyestuff type, the application route and depth of shade required due to the chemical reactions involved in the fixation of reactive dyes to fabrics [6]. This waste water derived from the textile industries needs suitable treatment(s) before its release in order to avoid dangerous increase of colour, pH, BOD and chemical oxygen demand (COD) in rivers or in drainage areas.

2.1.2 Techniques of Colour Removal

Effluents of textile industries are most commonly treated by physical or chemical processes. Some examples of the techniques most commonly used are listed in Table 2 with their associated pros and cons.

Increase in population, poverty and economic stagnation in many developing countries have left the society with economic problems and strategies for development which have increased the demand of water resources for agricultural, industrial and drinking water supply. Many countries promoted the reuse of treated waste water in order to minimize water consumption and to cause less pollution. Various treatment methods such as reduction, precipitation, coagulation and flocculation, flotation, adsorption on activated carbon, ion exchange, reverse osmosis and electro-dialysis are being also used in order to alleviate the problem of water pollution by dyes in the environment. Most of these methods used are expensive or ineffective, especially when the dye concentrations are in the order of 1–100 mg/L [10]. Hence, more effective and economical viable technologies are required for the colour removal from textiles effluent. Adsorption on activated carbon and other adsorbents from biomass is considered to be an efficient, promising and widely applicable fundamental approach in wastewater treatment processes contaminated with dyes. This approach (biosorption when adsorbent is

Table 2 Advantages and disadvantages of methods used for dye removal from an effluent

Techniques	Advantages	Disadvantages
Coagulation and flocculation	Effective for all dye	High cost
	Elimination of insoluble water dyes	High sludge production
Reverse osmosis	Removal of all reactive dyes, auxiliaries and minerals salts	A very high pressure is required
Nanofiltration	Removal of all types of dyes	A high investment cost is needed
	Easy to scale up	
	High effluent quality is produced	Influent must be treated
Ozonation	No sludge production	Highly expensive
	No alteration of volume is needed	Short half life
Electrochemical oxidation	Removal of small colloidal particles	Highly expensive
	No chemical use	Not effective for all dyes

derived from biomass) mainly hinges on its simplicity, economical viability, technical feasibility and socially acceptability [9].

2.2 Biosorption: A Green Remediation

Biosorption is the binding and concentration of adsorbate(s) from aqueous solutions (even very dilute ones) by certain types of inactive, dead, microbial biomass. The major advantages of biosorption over conventional treatment methods include: low cost, high efficiency, minimization of chemical or biological sludge, regeneration of biosorbents and possibility of metal recovery [16]. Another powerful technology is adsorption of heavy metals by activated carbon for treating domestic and industrial wastewater. However the high cost of activated carbon and its loss during the regeneration restricts its application. Since the 1990s the adsorption of heavy metals ions by low cost renewable organic materials has gained momentum. Recently attention has been diverted towards the biomaterials which are byproducts or the wastes from large scale industrial operations and agricultural waste materials.

Hence, pioneering research on biosorption of heavy metals, intrinsically guided by the emerging concept of *Green Chemistry*, has led to the identification of a number of microbial biomass types that are extremely effective in concentrating metals. *Green Chemistry* (environmentally benign chemistry) is the utilization of set of principles that reduces or eliminates the use or generation of hazardous substances in the design, manufacture and application of chemical products [8]. In practice, green chemistry is taken to cover a much broader range of issues than the definition suggests. As well as using and producing better chemicals with less waste, green chemistry also involves reducing other associated environmental impacts, including reduction in the amount of energy used in chemical processes.

Green Chemistry is not different from traditional chemistry in as much as it embraces the same creativity and innovation that has always been central to classical chemistry. However, there is a crucial difference in that historically synthetic chemists have not been seen to rank the environment very high in their priorities [8]. But with an increase in consciousness for environmental protection, environmental pollution prevention, safer industrial ecology and cleaner production technologies, throughout the world, there is a heightened interest and almost a grand challenge for chemists to develop new products, processes and services that achieve necessary social, economical and environmental objectives. Since the types of chemicals and the types of transformations are very much varied in the chemical industry and chemical research worlds, so are the green chemistry solutions that have been proposed.

Some types of biomass are waste byproducts of large-scale industrial fermentations while other metal-binding biomass types can be readily harvested from the oceans. Some biosorbents can bind and collect a wide range of heavy metals and organic molecules like dyes and pesticides with no specific priority, whereas others are specific for certain types of metals. When choosing the biomass for metal biosorption experiments, its origin is a major factor to be considered. In general terms, biomass can come from industrial wastes which should be obtained free of charge, organisms that can be obtained easily in large amounts in nature (e.g. bacteria, yeast, algae) or fast-growing organisms that are specifically cultivated or propagated for biosorption purposes (crab shells, seaweeds). Research on biosorption is revealing that it is sometimes a complex phenomenon where the pollutant species could be deposited in the solid biosorbent through various sorption processes, such as ion exchange, complexation, chelation, microprecipitation and oxidation/reduction. Biosorption equilibrium is achieved when the number of molecules leaving the surface of the adsorbent is equal to the number of molecules being adsorbed on the surface of the adsorbent. The equilibrium point is determined by parameters such as adsorbate dosage, pH, concentration of dye effluent and contact time.

Activated carbon was first reported for water treatment in the United States in 1930. Activated carbon is a crude form of graphite with a random or amorphous highly porous structure with a broad range of pore sizes having an extended area and crevices. Adsorption on granular activated carbons (GACs) is a proven, reliable technology for the advanced treatment of municipal and industrial wastewaters to remove small quantities of pollutants remaining in the wastewater following biological or physical–chemical treatments. Adsorption of metal ions present in dyes on carbon is more complex than uptake of organic compounds because ionic charges affect removal kinetics from solution. Despite carbon's prolific use to treat wastewater, the biggest barrier of its application by industries is its high-priced, requiring vast quantities of activated carbon and its regeneration. Hence, a growing exploitation to evaluate the feasibility and suitability of natural, renewable and low-cost materials are sought and the substitutes should be easily available, cheap and, above all, be readily regenerated, providing quantitative recovery. These are known as low cost activated carbons.

2.3 Adsorption Process

The adsorption process is carried out in various ways and is influenced by a number of parameters. The three most commonly methods used are batch process, fixed bed column and continuous process. Batch adsorption experiments can be carried out whereby a known mass of adsorbent is added to a known volume of dye solution. Gentle mixing is provided to the mixture for a certain period of time. At the end of the contact period, the solids are separated from the liquid by decantation or filtration and the liquid portion is then analysed to determine the residual final adsorbate concentration.

2.3.1 pH

pH of solution usually plays a major role in the adsorption process and it affects the solution chemistry of metals or dyes and the activity of the functional groups of the adsorbent. It is commonly agreed that the adsorption of dyes and metal cations such as that of cadmium, copper, zinc, lead, and others increases with increase in pH as the metal ionic species become less stable in the solution and they form precipitates that complicate adsorption process. Whereas in cases of dyes, different dye classes require different pH ranges. For instance, basic dyes require alkaline medium while reactive dyes demand strong acidic conditions for optimum dye adsorption. pH variation may also affect the state of active sites. When the binding groups are acidic, the availability of free sites depends on pH: at lower pH the active sites are protonated, therefore, competition between protons and metal ions for the sorption sites occurs. Complete desorption of the bound metal ions is possible, which is why acid treatment is a method for metal elution and regeneration of adsorbent material. A decrease in pH value by two units can in some cases result in approximately 90% reduction of metal binding. Extreme pH values may damage the structure of the adsorbent material. Microscopic observation have shown distorted cells, significant weight loss and decrease in the sorption capacity have also been observed.

2.3.2 Adsorbent Dosage

The adsorbent dosage also affects the adsorption process. In many instances, higher adsorbent dosages yield higher uptakes and higher percentage removal efficiency. Increase in adsorbent dosage generally increases the amount of solute adsorbed due to the increased in surface area which increases the number of binding sites.

2.3.3 Initial Dye Concentration

This parameter seems to have an impact on the adsorption process such that with higher initial concentration of dye solution, the uptake of dye is higher. This is because at lower initial concentration, the ratio of the initial number of moles of dyes to the available surface area is low. Hence, it is important to determine the maximum adsorption capacity of the adsorbent for which the experiments should be conducted at the highest possible initial dye concentration.

The batch column set up consists basically of a cylindrical column packed firmly over a certain depth with the adsorbent through which wastewater is allowed to flow. Initially, most of the solute will be adsorbed as it is exposed to the fresh adsorbent bed and thus, almost zero concentration would be expected at the column outlet [1]. However, for laboratory trails, the operation of the column should be terminated only when inlet solute concentration approximately equals to that at the outlet. This is because complete column saturation, which results in S-shaped breakthrough curve [18] is important to evaluate the characteristics and dynamic response of an adsorption column.

2.3.4 Bed Depth

An increasing bed depth leads to an increase in surface area of the adsorbent, thus providing a larger amount of available binding sites for adsorption. Therefore, leading to an increase in dyes or metal uptakes.

2.3.5 Initial Solute Concentration

The driving force for an adsorption process is the concentration difference between the solute on the adsorbent and that in the solution. Thus, an increased inlet solute concentration increases the concentration difference, which favors adsorption.

2.3.6 Mass of Adsorbent in Column

The accumulation of a solute in a fixed column is largely dependent on the amount of adsorbent loaded into the column. A large mass of adsorbent provides more available binding sites to the solute and so uptake of dyes or metals increases.

3 Experimental Methodology

3.1 Preparation of Chemically Activated Carbon from C. nucifera L.

Coconuts grow on tropical palm trees and are scientifically known as C. nucifera L. They are often sourced in geographical areas where coconuts are harvested including Malaysia, Sri Lanka and Philippines. Coconut shell is similar to hard woods in chemical composition but the lignin content is higher around 35–45% and the cellulose content is around 23–43% with an ash content of 3%. Coconut shell powder is preferred over the alternative materials available in the market such as bark powder, furfural and peanut shell powder because of its uniformity in quality and chemical composition, better properties in respect of water absorption and resistance to fungal attack. The adsorption process using CAC's involves various metal-binding mechanisms including ion exchange, surface adsorption, chemisorption, complexation, and adsorption and it finds a variety of applications in industry due to its several advantages over carbon made from other materials.

Yellow *C. nucifera* L. shells were obtained locally from the northern part of the island of Mauritius. The adsorbent preparation consisted of several stages such as cleaning of coconuts and removal of the shells from the fruits followed by an acid treatment in order to produce chemically activated carbon shells. A trial test was first carried out to determine the percentage yield of activated carbon obtained from the pretreatment of *C. nucifera* L. 200 dried *C. nucifera* L. were cleaned and the husks removed from them. Each was broken into two pieces and they were kept in the sun for drying for 3–5 days, after which the fruits were easily removed from the shells. The dried shells were then crushed into grade 2 sizes and about 8.95 kg of dried *C. nucifera* L. were obtained.

3.2 Acid Treatment of the C. nucifera L. shells

The dried pieces of shells of grade 2 (G2) sizes were mixed in a 1:1 weight ratio with concentrated sulphuric acid (24%). The shells were then allowed to soak in the acid for 24 h at room temperature. The samples were then filtered and the solid residue that was soaked coconut shells were placed in furnace and heated at 200°C where they were held for 24 h. After 24 h, the samples were removed from the furnace and allowed to cool back to room temperature. The samples were then washed with distilled water and soaked in 1% sodium hydrogen carbonate (NaHCO$_3$) solution to remove any remaining acid until pH of the activated carbon reached 6. The CAC's was then placed in the oven for drying at a temperature of 105°C for 3 h after which the samples were easily crushed using an electric grinder to be used for further adsorption experiments.

The crushed samples of the CAC's produced from the chemical treatment were sieved using sieves of three different mesh sizes of 15.000, 9.423 and 1.140 mm from which the particle size could be determined.

3.3 Preparation of Dye Solution

The RR158 dye was obtained from the Textile Department of Engineering of the University of Mauritius located at Réduit (Courtesy of Associate Professor Dr S. Rosunee and Dr N. Kistamah). Aqueous stock solution of 160 mg/L of the

Levafix brilliant red dye was then prepared using the RR158 dye powder and distilled water. Initial concentrations were adjusted to 120, 100, 80, 60, 40 and 20 mg/L to be used in the batch adsorption experiments.

3.4 Determination of Optimum Wavelength

About 20 mL of each solution prepared was pipetted into separate well dried spectrophotometer glass tubes and each tube was labeled respectively. A glass tube filled with distilled water was first placed in the spectrometer machine for zeroing the machine after which each tube was placed in the machine. The corresponding absorbance values were recorded for a range of wavelength from 400 to 600 nm for each dye solution. A graph of absorbance against wavelength for each con-centration of dye solutions was plotted from which the optimum wavelength was determined. Thus, the absorbance values of each solution of different concentra-tions (20–120 mg/L) were again read at the optimum wavelength value obtained and a second graph of absorbance against concentration of dye solution was plotted which contributed in determining a relationship between the absorbance values and concentrations of the solution.

A stock dye solution of 4500 mL of concentration of 100 mg/L was prepared using distilled water and the RR158 dye. A series of batch experiments were carried out using the dye solution prepared with a constant dosage of 7.000 ± 0.001 g/L of adsorbent of CAC's. The batch experiments were carried out at a constant agitation rate of 200 rpm at room temperature of 28°C but with varying pH values (2–9).

For each batch process, 500 mL of the dye solution was measured and transferred into a 500 mL conical flask. The pH of the dye solution was adjusted by using 1 M hydrochloric acid and 1 M sodium hydroxide solutions. The adsorbent was added to it and the opening of the conical flask was tightly closed with a rubber bung. The conical flask was then placed on the electronic stirrer and was run for 2 h. A sample of about 25 mL of the solution was pipetted at every time intervals of 20 min. The samples were filtered through the Whatman filter paper and the filtrates were poured in dif-ferent spectrophotometer tubes. The absorbance values of the samples were noted and recorded at the wavelength value and the corresponding residual dye concentration value could be obtained from the graph of absorbance against concentration of dye plotted (Absorbance $= 0.018 \times$ concentration of dye in solution). The efficiency of dye removal from the solution was then calculated at the different pH values.

3.5 Batch Adsorption Experiments

Initial batch adsorption experiments were carried out at the optimum pH value determined from the previous experiments at room temperature at an agitation rate of 200 rpm. The experiment was carried out firstly, with a range of constant doses

Table 3 Experimental conditions used for batch studies

Parameter	Initial dye concentration (mg/L)	Dosage of CAC's (g/L)	pH conditions and agitation rate (rpm)
Initial dye concentration (mg/L)	20, 40, 60, 80, 100, 120	7.5–20	pH 2 and 200
Dosage of adsorbent (g/L)	20–120	7.5, 10, 12.5, 15, 17.5, 20	pH 2 and 200

of adsorbent starting from 7.5 ± 0.001 g/L to 20 ± 0.001 g/L with dye solution of varying concentration of 20–120 mg/L and secondly, with constant concentration of dye solution and varying dosage of adsorbent. The batch experiments were run for 2–3 h. Samples were taken every 20 min and vacuum filtered through the Whatman filter paper and the filtrates collected in small spectrophotometer tubes were placed in the spectrophotometer whereby the absorbance values were read and noted, and the corresponding dye concentration values were obtained using calibration equation connecting absorbance and dye concentration. Thus, the residual dye concentrations of the solutions were calculated at each time interval and the effect of varying dosage and initial dye concentration analysed. Table 3 summarises the conditions of batch adsorption experiments.

3.6 Equilibrium Parameters of Adsorption

Equilibrium data which are generally known as the adsorption isotherms are the main requirements to understand the adsorption mechanism and it also aids in the process design. The traditional adsorption isotherm models such as Langmuir, Freundlich, Dubinin–Radushkevich (D–R) and Temkin isotherm model have been used to describe the equilibrium between RR158 and CAC's at constant temperature.

3.6.1 Langmuir Adsorption Isotherm Model

The Langmuir adsorption isotherm model is valid for monolayer adsorption onto a surface with a finite number of identical sites and is applicable in cases whereby there is a continual process of bombardment of molecules onto the surface and a corresponding evaporation of molecules from the surface to maintain zero rate of accumulation at the surface of equilibrium. The equation can be used only if the surface is homogeneous, that is adsorption energy is constant over all the sites and each site will accommodate only one molecule or atom. The Langmuir equation is as follows:

$$\frac{C_t}{q_t} = \frac{1}{q_{max}K_L} + \frac{C_t}{q_{max}}$$

where q_t is the dye concentration on the adsorbent (mg/g) at time t, C_t is the dye concentration in solution (mg/L) at time t, q_{max} is the monolayer adsorption capacity of the adsorbent(mg/g) and K_L is the Langmuir constant (L/mg) and is related to the free energy of adsorption.

The amount of RR158 dye adsorbed, q_t, in mg/g at time t was computed by using the following equation.

$$q_t = \frac{(C_0 - C_t)V}{m_s}$$

where C_0 is initial dye concentration (mg/L), C_t is residual dye concentration at time t, V is volume of solution (L) and m_s is mass of adsorbent used (g).

A plot of C_t/q_t versus C_t for the adsorption gives a straight line of slope $1/q_{max}$ and intercepts $1/q_{max} K_L$. The main feature of the Langmuir isotherm can be expressed by means of "R_L", a dimensionless constant referred to as separation factor or equilibrium parameter. R_L is calculated using equation below whereby C_0 is the initial dye concentration (mg/L) and K_L is a Langmuir parameter.

$$R_L = \frac{1}{1 + K_L C_0}$$

The related adsorption process is considered to be favorable in case R_L values lie between 0 and 1, and if $R_L = 0$, adsorption is considered to be linear and if R_L is greater than 1 adsorption process is unfavourable.

3.6.2 Freundlich Isotherm Model

This model is based on adsorption on a heterogeneous surface and is specified K_f and n are the Freundlich constants, characteristics of the system. The equation can be linearised in logarithmic form and Freundlich constants can be determined. The Freundlich isotherm is also more widely used but provides no information on the monolayer adsorption capacity, in contrast to the Langmuir model. A value of n between 2 and 10 shows good adsorption. The numerical value of $1/n$ less than 1 indicates that adsorption capacity is only slightly suppressed at lower equilibrium concentrations [14].

$$q_t = K_F C_t^n$$

where, q_t is the dye adsorbed per mass of adsorbent at time t (mg/g), C_t is the dye in the solution at time t (mg/L) and K_F ($mg^{1-1/n}L^{1/n}g^{-1}$) and n are the Freundlich model parameters.

3.6.3 Dubinin–Radushkevich (D–R) Isotherm Model

This model is applied to distinguish between physical and chemical adsorption. The linear equation of (D–R) isotherm model is as follows:

$$\text{Ln } q_t \equiv \text{Ln } q_m - \beta\varepsilon^2$$

where β is a constant connected with the mean free energy of adsorption per mole of the adsorbate (mol^2/kJ^2), q_m is the theoretical saturation capacity (mol/g), and ε is the Polanyi potential, which is equal to $RT \ln(1 + 1/C_t)$, where R (8.314 J/mol K) is the gas constant, and T (K) is the absolute temperature. Hence, by plotting ln q_t versus ε^2 it is possible to generate the value of q_m from the intercept, and the value of β from the slope.

3.6.4 Temkin Isotherm Model

This isotherm contains a factor that explicitly takes into account of adsorbent–adsorbate interactions. By ignoring the extremely low and large value of concentrations, the model assumes that heat of adsorption of all molecules in the layer would decrease linearly rather than logarithmic with coverage. As implied in the equation below, its derivation is characterized by a uniform distribution of binding energies. Temkin equation is excellent for predicting the gas phase equilibrium (when organization in a tightly packed structure with identical orientation is not necessary), conversely complex adsorption systems including the liquid-phase adsorption isotherms are usually not appropriate to be represented.

$$q_t = \frac{R_T}{b_T \text{ Ln } A_T C_t}$$

whereby, q_t is the amount of dye adsorbed per mass of adsorbent (mg/g), R is the gas constant and equal to 8.314 kJ/K mol and T is temperature in Kelvin. A_T and b_T are the Temkin parameters while C_t is the residual dye concentration at time t.

3.7 Kinetic Modelling

3.7.1 Pseudo First-Order Equation

Lagergren (1898) expressed the equation based on solid capacity and it is generally expressed as follows:

$$\frac{dq}{dt} = k_{1,ad}\left(q_{eq} - q\right)$$

where q_{eq} and q_t are the adsorption capacity at equilibrium and at time t, respectively (mg/g), $k_{1,ad}$ is the rate constant of pseudo first-order adsorption (L/min). After integration and applying boundary conditions, $t = 0$ to $t = t$ and $q = 0$ to $q = q_{eq}$; the integrated form of equation becomes

$$Log(q_{eq} - q) = Log\, q_{eq} - \frac{k_{1,ad}}{2.303}$$

A graph of $log(q_{eq} - q)$ against $log\, q_{eq}$ helps to predict the rate constant of first order through its slope and y-intercept values. The theoretical q_{eq} values determined were compared with experimental values.

3.7.2 Pseudo Second-Order Equation

If the rate of sorption is a second-order mechanism, the pseudo second-order chemisorption kinetic rate equation is expressed as:

$$\frac{dq_t}{dt} = k_{2,ad}(q_{eq} - q_t)^2$$

where $k_{2,ad}$ is the rate constant of second-order adsorption (g/mg min). For the boundary conditions $t = 0$ to $t = t$ and $q = 0$ to $q = q_{eq}$, the latter equation may be integrated and rearranged in a linear form that is expressed as follows:

$$\frac{t}{q} = \frac{1}{k_{2,ad}q_{eq}^2} + \frac{1}{q_{eq}}t$$

If the initial adsorption rate is h (mg/g min) and is equal to $k_2 q_{eq}^2$ then, further simplification gives:

$$\frac{t}{q_t} = \frac{1}{h} + \frac{1}{q_e}t$$

The plot of (t/q_t) verses t of equation gives a linear relationship from which q_e and K_2 can be determined.

3.7.3 Intra-Particle Diffusion Model

The intra-particle diffusion model is expressed as:

$$R(\%) = k_{id}(t)^a$$

where R is the percentage of dye adsorbed, t is the contact time (h), a is the gradient of linear plots, k_{id} is the intraparticle diffusion rate constant ($h - 1$), and it depicts the adsorption mechanism while k_{id} may be taken as a rate factor that is, percent dye adsorbed per unit time [15]. A linearised form of the equation is followed by:

$$Log\, R = Log\, k_{id} + a\, Log(t)$$

Larger k_{id} values illustrate a better adsorption mechanism, which is related to an improved bonding between RR158 and the adsorbent particles [4].

3.8 Batch Column Experiments

The column tests were carried out in a glass column with a diameter of 75.0 mm and length of 18 cm. The column was indiscriminately filled with CAC's samples and fixed in a vertical position. It was noticed that there was a decrease in pressure in the column and a decrease in the surface area available for mass transfer between the adsorbent and adsorbate due to the formation of air pockets. Thus, deionised water was used to wash the CAC's in order to remove air bubbles. The column filled with a known volume of dried CAC's was immersed in a glass beaker containing deionised water and sucked for about 15 min with a vacuum pump to eliminate the air pockets formed in the column [1].

A stock solution of the RR158 dye of 120 mg/L was prepared and different dye concentrations of 60, 80, 100, and 120 mg/L were prepared. The pH of the dye solution was adjusted to the optimum value with the use of 1 M hydrochloric acid and 1 M sodium hydroxide solution. The dye solution was then poured at the top of the column and allowed to flow down the column through gravity. Samples of the solution were taken from the bottom of the column every 20 min and analysed in the spectrophotometer to find out the residual dye concentration in the solution. Thus, the amount of dye adsorbed from the solution was obtained using the equation below:

$$R(\%) = \frac{C_0 - C_F}{C_0} \times 100$$

C_0 and C_t are the dye concentrations in mg/L initially and at a given time t, respectively. V is the volume of the dye solutions in mL and R is the percentage of dye removed from solution. Further experiments studying the effect of initial dye concentration, initial volume of solution and height of the column were also considered. The details of experimental conditions of the fixed bed column studies of RR158 dye onto CAC's for the entire column runs undertaken are listed in Table 4.

3.8.1 Thomas Model

Thomas derived the mathematical expression for a column with a typical break-through curve that is expressed in the equation below, where C_0 is the initial dye concentration (mg/L), C_t is the equilibrium concentration (mg/L) at time t (min),

Table 4 Experimental conditions used for batch column studies

Parameter	Initial dye concentration (mg/L)	Initial volume of dye (mL)	Bed height of column (cm)
Initial dye concentration (mg/L)	60, 80, 100, 120	150	6
Initial volume of dye (mL)	60	100, 150, 200, 250	6
Bed height of column (cm)	120	250	6, 8, 10, 12

k_T is the Thomas constant (L/min mg), F is the volumetric flow rate (L/min), q_0 is the maximum column adsorption capacity (mg/g), m is the mass of adsorbent (g) and V is the throughput volume (L).

$$\frac{C_t}{C_0} = \frac{1}{1 + e^{[k_T(q_0 m - C_0 V)/F]}}$$

Hence, a plot of $\ln(C_0/C_t - 1)$ versus V as stated in immediately below 18 gives a straight line with a slope of $(-k_T C_0/F)$ and an intercept of $(k_T q_0 m/F)$. Therefore, k_T and q_0 can be obtained.

$$\mathrm{Ln}(C_0/C_t - 1) = \frac{k_T q_0 m}{F} - \frac{k_T C_0 V}{F}$$

3.8.2 Data Analysis

Data analysis also called error analysis is important in order to confirm the fit model for the adsorption system with the experimental data obtained. Data analysis was carried out for Langmuir, D–R model, pseudo first order and second order as well as for Thomas model. The calculated expressions of the error functions used were as follows:

The residual sum of errors (RSS):

$$\mathrm{RSS} = \sum (q_c - q_e)^2$$

The sum of the absolute errors (SAE):

$$\mathrm{SAE} = \sum |q_c - q_e|$$

The average relative error (ARE):

$$\mathrm{ARE} = \frac{[\sum |(q_c - q_e)/q_e|]}{n}$$

where n is the number of experimental data points, q_c is the predicted (calculated) quantity of RR158 adsorbed onto CAC's according to the isotherm equations and q_e is the experimental data.

Table 5 Characteristics of the CAC's

Characteristics	Value
Particle size (mm)	≤ 1.14
Bulk density (g/L)	795.00 ± 14.34
Ash content (%)	4.478 ± 0.211
Moisture content (%)	4.335 ± 0.237
pH	6.00 ± 0.01

4   Results and Discussions

4.1   Yield of CAC's

The yield of pretreated coconut shell obtained was $89.378 \pm 0.027\%$ indicating that there was little loss of active biomass of the *C. nucifera* L. shells and the low value of standard deviation implies that the adsorbent preparation experiments were well carried out with minimum losses of coconut shells.

4.2   Characteristics of the Adsorbent

The characteristics of the adsorbent are given in Table 5.

The chemically activated carbon coconut shell had a low moisture content and ash content.

4.3   Optimum Wavelength

A series of various dye concentration were tested in the spectrophotometer at varying wavelength values and a graph of absorbance verses wavelength was plotted from which the optimum wavelength was 520 nm. Also the graph obtained had high correlation coefficients ($0.997 < R^2 < 1$). A calibrated curve of dye concentration versus absorbance values was plotted as shown in Fig. 2 which showed that the spectrophotometer reading was proportional to the concentration of the dye solution such that absorbance value of a dye solution was 0.018 times its concentration. This calibration curve was then used to determine the concentration of dye in the permeate solution collected from time to time through-out the 2–3 h experiment.

4.4   Optimum pH

Adsorption experiments were performed over a range of pH starting from 2.00 ± 0.01 to 9.00 ± 0.01. Figure 3 shows that the adsorption of RR158 was highest at pH 2.00 ± 0.01 which was $51.51 \pm 10.60\%$ and as pH increased, the dye

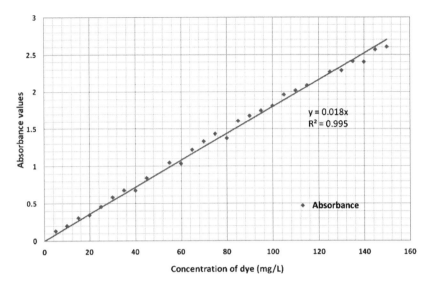

Fig. 2 Graph of absorbance against concentration of dye

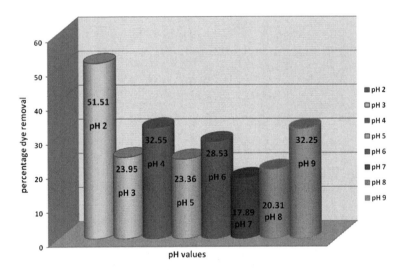

Fig. 3 Dye removal at the corresponding pH values

removal was in a range of 17.00 ± 10.60 to $33.00 \pm 10.60\%$. Hence, it was observed that adsorption of dyes is highly pH dependent, which influences the adsorption of both anions and cations at the solid–liquid interface. The effect of pH on the adsorption capacity can be attributed to the chemical forms in the solution at a specific pH and also due to different functional groups on the adsorbent surface, which become active sites for the metal biding at a specific pH. Therefore, an increase in pH may cause either an increase or a decrease in the adsorption capacity, resulting

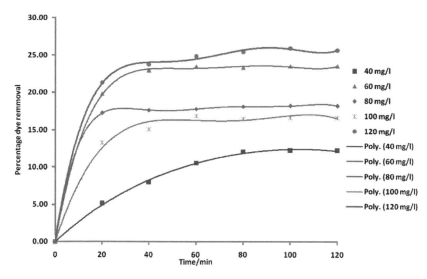

Fig. 4 Graph of removal of dye with respect to time. Experimental conditions: at room temperature and pressure and 200 rpm, pH: 2 and dosage of adsorbent: 7.5 g/L

in different optimum pH values dependent on the type of the adsorbent. The increase in the dye adsorbed at lower pH 2 could be well explained by protonation properties of the adsorbent. At low pH the negative charges at the surface of internal pores are neutralized and some more new adsorption sites are developed and as a result the surface provided a positive charge for the RR158 dye to get adsorbed thus increasing the uptake of dye. Above pH 7 the RR158 dye particles acquire a negative surface charge leading to a lesser dye uptake since dye molecule becomes neutral at that pH value. Also, in alkaline medium, precipitation tends to appear and interfere with the accumulation or adsorbent deterioration as the dyes particles tend to flocculate and formed a cake layer thus reducing adsorption efficiency [12].

4.5 Batch Studies: Removal of RR158 with Time

The removal of RR158 dye from the synthetic solution with respect to time for each solution at different initial values of dye concentration is illustrated in Figs. 4, 5, 6, 7, 8 and 9. It was observed that at low dosages of adsorbent, the dye removal decreased as the initial dye concentration rose. For instance at dosage of 7.5 g/L, the maximum dye uptake of the 40 mg/L dye solution was about 12.00% while that of a 100 mg/L dye solution was 24.00%. As dosage of adsorbent and initial dye concentration increases, the dye uptake also increased. As far as the equilibrium time taken to achieve maximum dye uptake was concerned, it was seen that at low adsorbent dosages, high concentrations of RR158 dye solutions took longer to reach equilibrium while at high dosages (15, 17.5 and 20 g/L), low dye

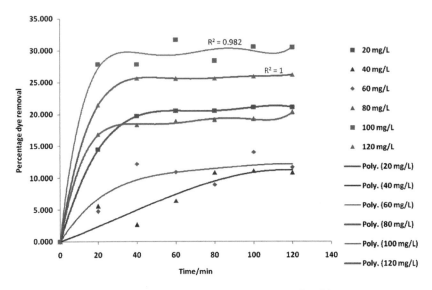

Fig. 5 Graph of percentage removal of dye with respect to time. Conditions: room temperature and 200 rpm, pH: 2 and dosage of adsorbent: 10 g/L

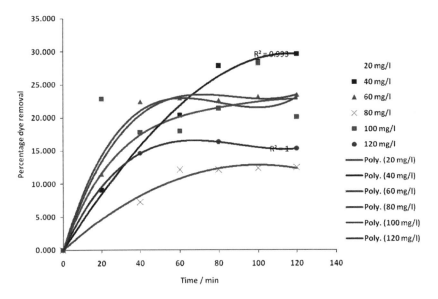

Fig. 6 Graph of removal of dye with respect to time. Conditions: at rtp and 200 rpm, pH: 2 and dosage of adsorbent: 12.5 g/L

concentration solutions (20 and 40 mg/L) had lower equilibrium times. The equilibrium time achieved for varying initial dye concentrations at various dosages of adsorbent determined from the plotted graphs are tabulated in Table 6.

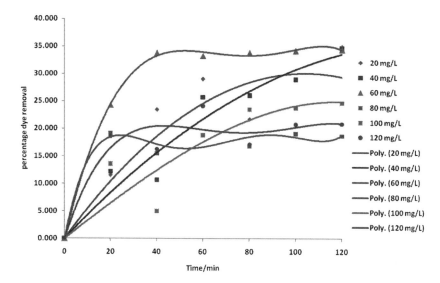

Fig. 7 Graph of removal of dye with respect to time. Conditions: at rtp and 200 rpm, pH: 2 and dosage of adsorbent: 15 g/L

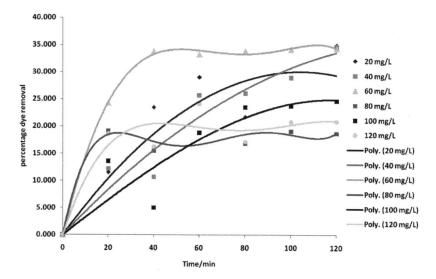

Fig. 8 Graph of removal of dye with respect to time. Conditions: at room temperature and 200 rpm, pH: 2 and dosage of adsorbent: 17.5 g/L

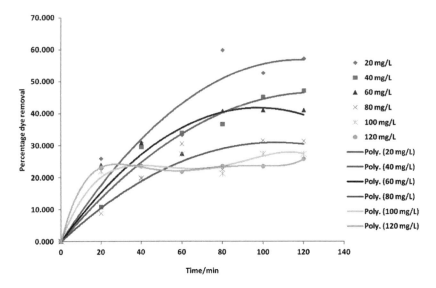

Fig. 9 Graph of percentage removal of dye with respect to time. Conditions: at room temperature and 200 rpm, pH: 2 and dosage of adsorbent: 20 g/L

Table 6 Equilibrium time for the respective batch experiments

Dye concentration (mg/L)	Equilibrium time, t_{eq} (min)					
	Adsorbent dosage					
	7.5 (g/L)	10 (g/L)	12.5 (g/L)	15 (g/L)	17.5 (g/L)	20 (g/L)
20	90	40	120	100	100	120
40	80	100	120	120	120	120
60	40	60	60	40	40	100
80	40	40	80	120	40	100
100	40	40	60	120	40	40
120	40	40	60	40	40	40

4.6 Effect of Dye Concentration on Batch Adsorption Process

In the present work, the initial RR158 concentration was varied from 40 to 120 mg/L while maintaining the adsorbent dosage at 7.5, 10, 12.5, 15, 17.5 and 20 g/L. In batch adsorption, the initial concentration of the dye molecules in the solution plays a key role as a driving force to overcome the mass transfer resistance between the aqueous and solid phases. Therefore, the amount of dye particles adsorbed is expected to be higher with a higher initial concentration [3]. This was indeed the case (Fig. 10) for dosages of CAC's of 7.500 and 10.000 ± 0.001 g/L that showed an increase in removal of RR158 as the initial dye concentration increased from 40 to 80 mg/L until an optimum value was reached. However, the dye removal decreased from 57.18 to 36.44% when the dye concentration

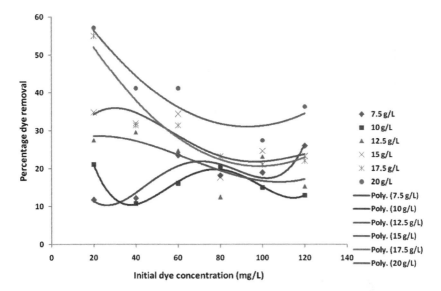

Fig. 10 Dye removal with initial dye concentration variation at constant dosages of adsorbent. Experimental conditions: room temperature, agitation: 200 rpm

increased from 40 to 120 mg/L with the same contact time and adsorption temperature for a constant dosage of 20 g/L of CAC's while, for a dosage of 17.500 g/L the dye removal decreased from 55.00 to 22.08% and from 27.5 to 15.29% when the dosage was 12.500 ± 0.001 g/L. This decrease in dye uptake with an increase in initial dye concentration may be due to an increase in the number of RR158 particles for the fixed amount of adsorbent and may be due to the complete utilization of adsorption surface and active sites available which was not possible in low concentration [2].

4.7 Effect of Dosages of CAC's on RR158 Dye Removal

The effect of CAC's (varying from 7.500 to 20.000 g/L) on the RR158 dye removal efficiency of six different initial dye concentrations of 20, 40, 60, 80, 100 and 120 mg/L is presented in Fig. 11 which shows that the removal of RR158 increased with an increase in the adsorbent dosage. However, once almost all RR158 was adsorbed, the contribution of additional CAC's was found to be insignificant. The dye removal increased from 11.75 ± 10.60 to 57.18 ± 10.60% when there was an increase in the adsorbent dosage from 7.5 ± 0.001 to 20 ± 0.001 g/L for dye solution of 20 mg/L, and from 12.18 to 47.11% at constant dye concentration of 40 mg/L. The rapid increase in adsorption with increasing adsorbent dose can be attributed to greater surface area and availability of more adsorption sites. Thus, the increase in colour removal at higher doses of

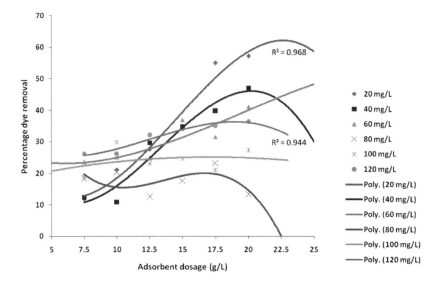

Fig. 11 Graph of percentage dye removal with respect to adsorbent dosage. Experimental conditions: room temperature, agitation: 200 rpm

adsorbent could be due to the very fast superficial adsorption onto the CAC's surface that produced a lower solute concentration in the solution than when the adsorbent dose was lower (Garg 2008). The adsorption rate increased slowly after the critical dose of 7.500 g/L for dye solution of 60, 100 and 120 mg/L. However, the adsorption efficiency dropped from $20.00 \pm 10.60\%$ to $12.48 \pm 10.60\%$ when the adsorbent dosage rose from 7.500 ± 0.001 to 12.500 ± 0.001 g/L for the dye solution of 80 mg/L. The drop in the dye removal could basically be due to the sites remaining unsaturated during the adsorption process. The decrease in dye uptake value was due to the splitting effect of flux (concentration gradient) between the adsorbate and the adsorbent.

4.8 Equilibrium Isotherms for Batch Experiments

4.8.1 Langmuir Model

The experimental data were fitted to the Langmuir model and the linearised forms are shown in Figs. 12, 13, 14, 15, 16 and 17 which had high correlation coefficients ($R^2 > 0.85$), thus, showing very good agreement with this model. The high correlation coefficients also indicated a good agreement between the parameters and confirmed the monolayer adsorption of RR158 onto the adsorbent surface. A summary of the results for the calculated values of the Langmuir parameters and the R_2 values is tabulated in Table 7. R_L indicates the isotherm shape according to

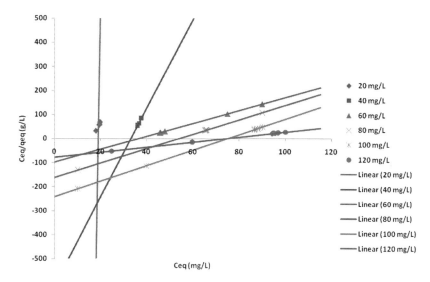

Fig. 12 The linearised Langmuir isotherm for RR158 on CAC's. Experimental conditions: room temperature, 200 rpm, adsorbent dosage = 7.5 g/L

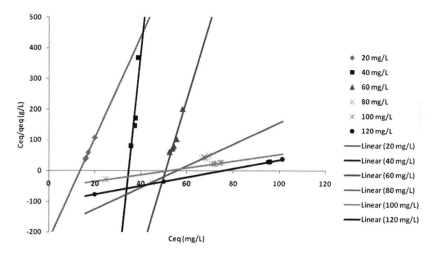

Fig. 13 The linearised Langmuir isotherm for RR158 on CAC's. Experimental conditions: at room temperature, 200 rpm, adsorbent dosage = 10 g/L

the following adsorption characteristics, $R_L > 1$ (is unfavourable), $R_L = 1$ (linear adsorption), $R_L = 0$ (is irreversible) and $0 < R_L < 1$ (is favourable). The R_L values for the present study were found to be 0.264–0.488 for initial dye concentrations of 20–120 mg/L, which is consistent with the requirement for a favourable adsorption

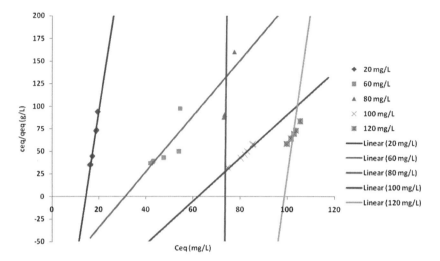

Fig. 14 The linearised Langmuir isotherm for RR158 on CAC's. Experimental conditions: at room temperature, 200 rpm, adsorbent dosage $= 12.5$ g/L

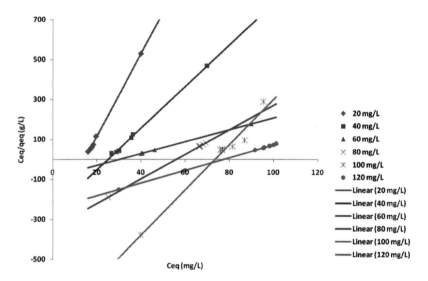

Fig. 15 The linearised Langmuir isotherm for RR158 on CAC's. Experimental conditions: room temperature, 200 rpm, adsorbent dosage $= 15$ g/L

process. High Langmuir constant K_L indicates high affinity for the binding of RR158. The values of q_{max} obtained were in the range of 0.0017–0.9728 mg/g for the synthetic dye solutions of 20–120 mg/L and the K_L values obtained were in the range of 0.01–0.05 L/mg.

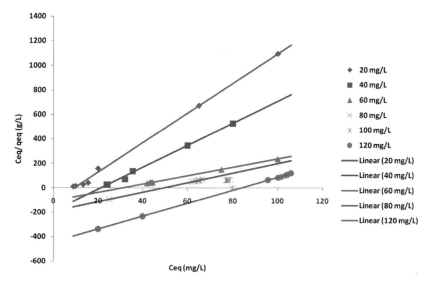

Fig. 16 The linearised Langmuir isotherm for RR158 on CAC's. Experimental conditions: at room temperature, 200 rpm, adsorbent dosage = 17.5 g/L

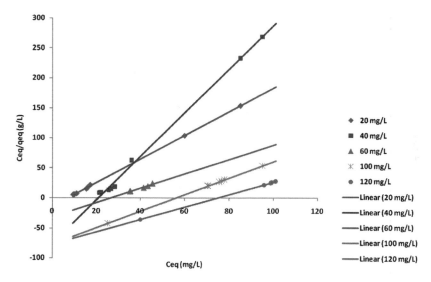

Fig. 17 The linearised Langmuir isotherm for RR158 on CAC's. Experimental conditions: at room temperature, 200 rpm, adsorbent dosage = 20 g/L

4.8.2 Freundlich Model

The values of the Freundlich constants determined from the graphs displayed in Figs. 18, 19, 20, 21, 22 and 23 are given in Table 7. The high correlation

Table 7 Langmuir and Freundlich constants of RR158 adsorption on CAC's

Dye concentration (mg/L)	Langmuir				Freundlich			
	R^2	q_{max} (mg/g)	K_L (L/mg)	R_L (–)	R^2	n	$1/n$	K_f
Adsorbent dosage: 7.5 g/L								
20	0.990	0.0017	0.0525	0.488	0.827	20.28	0.049	3.76E+25
40	0.990	0.0543	0.0301	0.454	0.997	9.065	0.110	8.81E+13
60	1.000	0.3719	0.0271	0.380	0.999	3.573	0.280	1.62E+06
80	1.000	0.3349	0.0185	0.404	0.999	4.573	0.219	3.89E+08
100	1.000	0.3100	0.0133	0.429	0.998	5.621	0.178	1.79E+11
120	1.000	0.9381	0.0134	0.383	0.998	3.233	0.309	1.06E+07
Adsorbent dosage: 10 g/L								
20	0.999	0.0604	0.0738	0.404	0.999	4.702	0.213	1.82E+05
40	0.841	0.0139	0.0288	0.464	0.953	14.37	0.070	8.99E+21
60	0.922	0.0396	0.0197	0.459	0.973	10.81	0.093	3.76E+18
80	0.999	0.2838	0.0180	0.410	0.998	4.456	0.224	2.29E+08
100	1.000	0.9074	0.0194	0.340	0.999	2.406	0.416	9.39E+04
120	0.999	0.7168	0.0133	0.385	0.999	3.233	0.309	8.45E+06
Adsorbent dosage: 12.5 g/L								
20	0.966	0.0595	0.0693	0.419	0.998	4.295	0.233	7.76E+04
60	0.564	0.3227	0.0321	0.342	–	–	–	–
80	0.844	0.0030	0.0136	0.479	0.915	26.31	0.038	1.08E+49
100	0.977	0.4172	0.0161	0.383	0.990	3.419	0.292	6.08E+06
120	0.945	0.0538	0.0101	0.451	0.972	10.8	0.093	7.89E+21
Adsorbent dosage: 15 g/L								
20	0.988	0.0481	0.0685	0.422	0.964	4.297	0.233	6.62E+04
40	0.997	0.0958	0.0398	0.386	0.958	3.823	0.262	2.74E+05
60	1.000	0.3361	0.0327	0.337	0.999	2.445	0.409	1.17E+04
80	1.000	0.1627	0.0178	0.412	0.998	4.814	0.208	6.17E+08
100	0.975	0.0884	0.0136	0.424	0.536	3.339	0.299	2.93E+06
120	0.999	0.3141	0.0129	0.392	0.983	3.820	0.262	6.00E+07
Adsorbent dosage: 17.5 g/L								
20	0.997	0.0830	0.1075	0.317	0.885	2.044	0.489	79
40	0.966	0.1112	0.0467	0.349	0.977	3.079	0.325	1.66E+04
60	1.000	0.2931	0.0325	0.339	0.998	2.446	0.409	1.02E+04
80	1.000	0.2558	0.0203	0.382	0.997	3.297	0.303	1.01E+06
100	0.999	0.2348	0.0160	0.384	0.999	3.854	0.259	2.29E+07
120	0.999	0.1893	0.0119	0.412	0.990	4.934	0.203	8.82E+09
Adsorbent dosage: 20 g/L								
20	0.996	0.5076	0.1391	0.264	0.982	1.330	0.752	38
40	0.999	0.2759	0.0478	0.343	0.916	2.741	0.365	1.28E+04
60	0.983	0.8389	0.0377	0.306	0.902	2.130	0.469	6.87E+03
80	0.999	0.3266	0.0191	0.396	0.950	4.208	0.238	8.52E+07
100	0.999	0.7278	0.0178	0.359	0.998	3.204	0.312	2.96E+06
120	1.000	0.9728	0.0133	0.385	0.999	3.202	0.312	9.82E+06

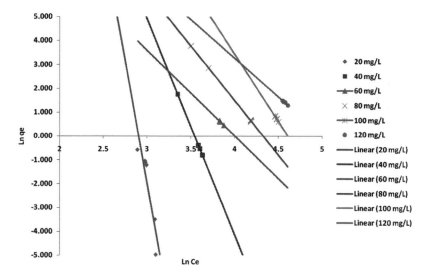

Fig. 18 The linearised Freundlich isotherm for RR158 on CAC's. Experimental conditions: at room temperature, 200 rpm, adsorbent dosage = 7.5 g/L

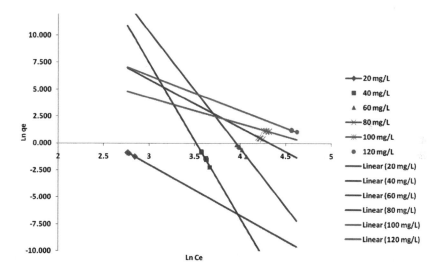

Fig. 19 The linearised Freundlich isotherm for RR158 on CAC's. Experimental conditions: at room temperature, 200 rpm, adsorbent dosage = 10 g/L

coefficients of the linearised Freundlich models obtained imply that heterogeneous surface conditions may co-exist within the monolayer adsorption under the applied experimental conditions. Hence, the overall adsorption of RR158 on the biomass is complex, involving more than one mechanism, such as ion exchange, surface complexation and electrostatic attraction [5].

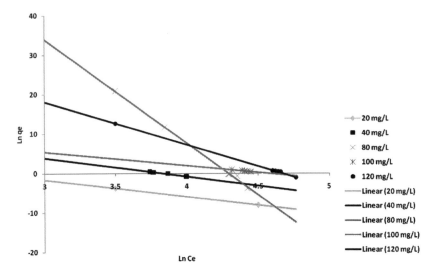

Fig. 20 The linearised Freundlich isotherm for RR158 on CAC's. Experimental conditions: at room temperature, 200 rpm, adsorbent dosage = 12.5 g/L

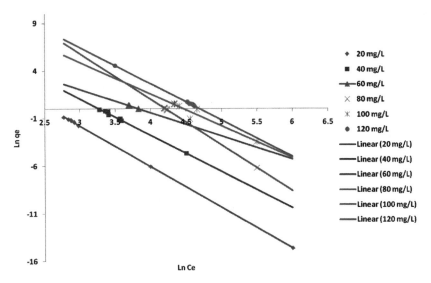

Fig. 21 The linearised Freundlich isotherm for RR158 on CAC's. Experimental conditions: room temperature, 200 rpm, adsorbent dosage = 15 g/L

The constants of the Freundlich isotherm, K_F were in the range of 79 to 3.76 E+21. High values of these parameters at pH 2 showed easy separation of RR158 from the aqueous solution [17]. The values of the Freundlich constants showed a relatively easy uptake of RR1528 with high adsorptive capacity of CAC's.

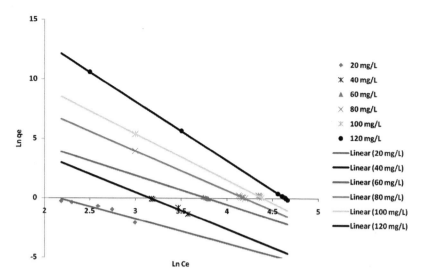

Fig. 22 The linearised Freundlich isotherm for RR158 on CAC's. Experimental conditions: room temperature, 200 rpm, adsorbent dosage = 17.5 g/L

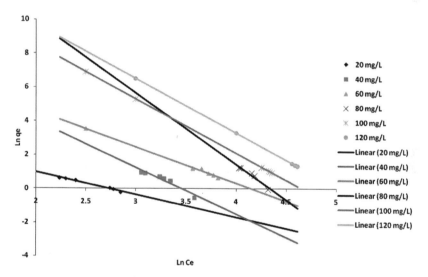

Fig. 23 The linearised Freundlich isotherm for RR158 on CAC's. Experimental conditions: room temperature, 200 rpm, adsorbent dosage = 20 g/L

The values of n for the present work were in the range of 2–10 hence indicating that adsorption is beneficial and the values of $1/n$ less than 1 generally indicated that adsorption capacity was only slightly suppressed at lower equilibrium concentrations. However, there were some exceptions such that at constant dosage of CAC's and a dye solution of 20 mg/L, n was 20.28.

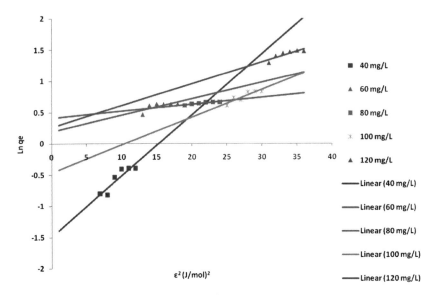

Fig. 24 The linearised D–R isotherm for RR158 on CAC's. Experimental conditions: room temperature, 200 rpm, adsorbent dosage $= 7.5$ g/L

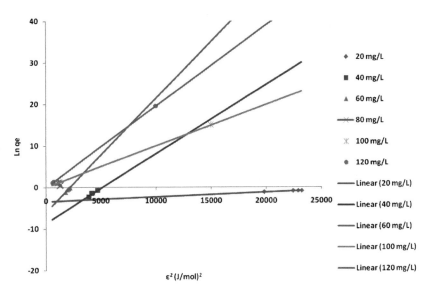

Fig. 25 The linearised D–R isotherm for RR158 on CAC's. Experimental conditions: at room temperature, 200 rpm, adsorbent dosage $= 10$ g/L

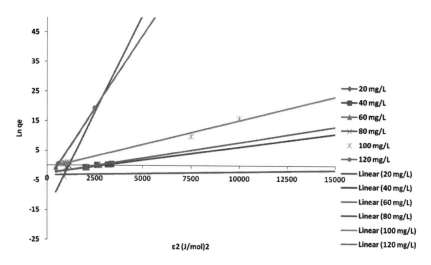

Fig. 26 The linearised D–R isotherm for RR158 on CAC's. Experimental conditions: room temperature, 200 rpm, adsorbent dosage = 12.5 g/L

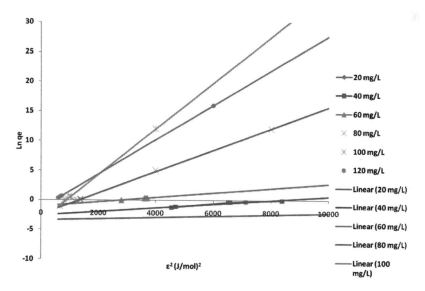

Fig. 27 The linearised D–R isotherm for RR158 on CAC's. Experimental conditions: room temperature, 200 rpm, adsorbent dosage = 15 g/L

4.8.3 Dubinin–Radushkevich Isotherm Model

The values for the theoretical saturation capacity and that of β which is a constant connected with the mean free energy of adsorption per mole of adsorbate were determined from the slopes and y intercepts of the linearly regressed lines

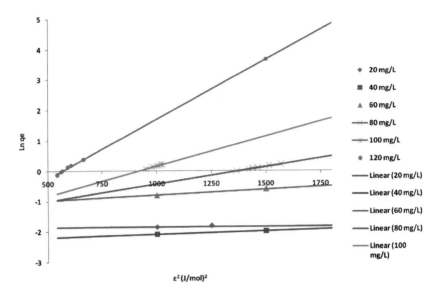

Fig. 28 The linearised D–R isotherm for RR158 on CAC's. Experimental conditions: room temperature, 200 rpm, adsorbent dosage = 17.5 g/L

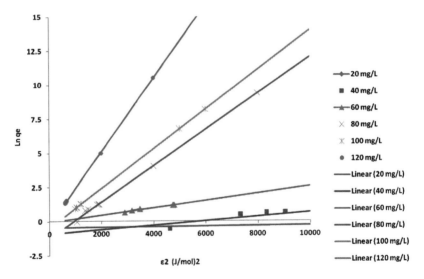

Fig. 29 The linearised D–R isotherm for RR158 on CAC's. Experimental conditions: room temperature, 200 rpm, adsorbent dosage = 20 g/L

illustrated in Figs. 24, 25, 26, 27, 28 and 29. A summary of the results obtained is tabulated in Table 8. The R^2 values were greater than 0.945 for varying initial concentration of dye solution for fixed amount of adsorbent of 10 g/L, greater than 0.899 for adsorbent dosage of 12.5 g/L, greater than 0.916 for adsorbent dosage of

Table 8 Summary of results obtained for values of Dubinin–Radushkevich constants

Dye concentration (mg/L)	R^2 values	k_{ad}	q_s (mg/g)	R^2 values	k_{ad}	q_s (mg/g)
	Adsorbent dosage: 7.5 g/L			Adsorbent dosage: 15 g/L		
20	0.004	0.06700	0.1579	0.934	−0.00012	0.0280
40	0.827	−0.09700	0.2245	0.936	−0.00019	0.2365
60	0.524	−0.02600	1.2105	0.998	−0.00038	0.3423
80	0.890	−0.01100	1.5008	0.997	−0.00100	0.0850
100	0.848	−0.04400	0.6263	0.916	−0.00300	0.0325
120	0.769	−0.03400	1.3034	0.988	−0.00300	0.2131
	Adsorbent dosage: 10 g/L			Adsorbent dosage: 17.5 g/L		
20	0.997	−0.00011	0.0316	0.900	−0.00004	0.0692
40	0.945	−0.00100	0.0002	0.958	−0.00023	0.0872
60	0.961	−0.00200	0.0024	0.997	−0.00046	0.2302
80	0.997	−0.00100	0.1673	0.994	−0.00100	0.2101
100	0.998	−0.00100	0.9305	0.998	−0.00200	0.1643
120	0.998	−0.00200	0.6011	0.984	−0.00400	0.0976
	Adsorbent dosage: 12.5 g/L			Adsorbent dosage: 20 g/L		
20	0.990	−0.00013	0.0301	0.945	−0.00002	0.6213
40	0.966	−0.00093	0.0660	0.803	−0.00017	0.2539
60	0.998	−0.00100	0.0630	0.982	−0.00028	0.8411
80	0.899	−0.01300	0.0000	0.909	−0.00100	0.2587
100	0.992	−0.00100	0.3420	0.996	−0.00100	0.6163
120	0.999	−0.01000	0.0035	0.995	−0.00200	0.8163

15 g/L, higher than 0.900 for adsorbent dosage of 17.5 g/L and higher than 0.909 for adsorbent dosage of 20 g/L. While for adsorbent dosage of 7.5 g/L, the R^2 values varied between 0.004 and 0.890 indicating that the D–R model fitted well at high dosages of adsorbent, it was not a good model at low dosages of adsorbent for the adsorption process. The values of ε^2 were in the range of 11.91 and 17.96 kJ/mol for dye concentration of 20 mg/L with 7.5 g/L of adsorbent. For an adsorption process with ε^2 between 8 and 16 kJ/mol, the process is known to follow chemical ion exchange while the process is likely to be considered as physical adsorption with ε^2 less than 8 kJ/mol. Therefore, the adsorption of RR158 by CAC's in the present study was considered to be influenced physically for initial dye concentration of 40, 60, 80, 100 and 120 mg/L as ε^2 values were below 8 kJ/mol whereas at lower initial concentrations of 20 mg/L, the adsorption process was influenced by chemisorption.

4.8.4 Temkin Isotherm Model

The values of q_e versus $\ln C_e$ were plotted so that linearised form of Temkin model is obtained in the form of $q_e = A + B \ln C_e$, whereby A is the y intercept and B is the gradient. The linearised Temkin model are illustrated in Figs. 30, 31, 32, 33, 34 and 35 followed by a summary Table 9 that gives the calculated values of the Temkin parameters. The R^2 values obtained from the linear method were above

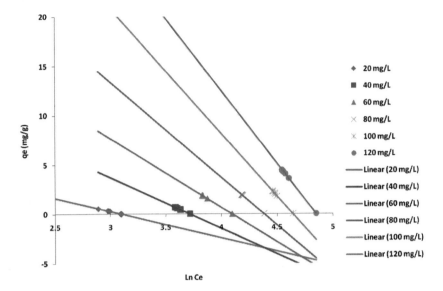

Fig. 30 The linearized Temkin model of RR158 by CAC's. Experimental conditions: room temperature; agitation rate: 200 rpm; biomass dosage: 7.5 g/L; pH: 2

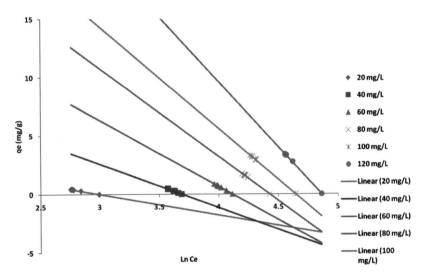

Fig. 31 The linearized Temkin model of RR158 by CAC's. Experimental conditions: room temperature; agitation rate: 200 rpm; biomass dosage: 10 g/L; pH: 2

0.980 which indicated that the Temkin model was well fitted to the experimental data. The parameters of b_T increased with increasing initial dye concentration from 20 mg/L to 120 mg/L. The values of b_T were in the range of -917.96 to -168.43 at fixed adsorbent dosage of 7.5 g/L, -1394.24 to -222.01 for adsorbent dosage

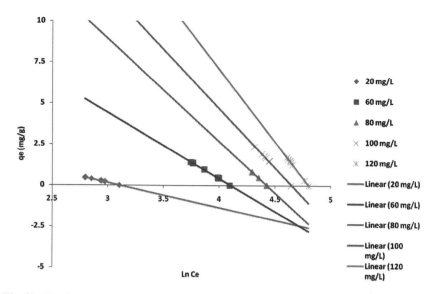

Fig. 32 The linearized Temkin model of RR158 by CAC's. Experimental conditions: room temperature; agitation rate: 200 rpm; biomass dosage: 12.5 g/L; pH: 2

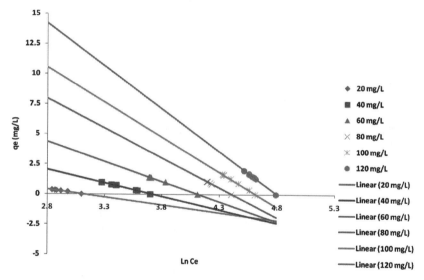

Fig. 33 The linearized Temkin model of RR158 by CAC's. Experimental conditions: room temperature; agitation rate: 200 rpm; biomass dosage: 15 g/L; pH: 2

of 10 g/L, -1637.52 to -280.62, -1949.31 to -348.71, -3093.01 to -398.71 and -1266.01 to -164.62 for adsorbent dosages of 12.5, 15, 17.5 and 20 g/L respectively. It was also observed that the b_T values lied in a wider range as the

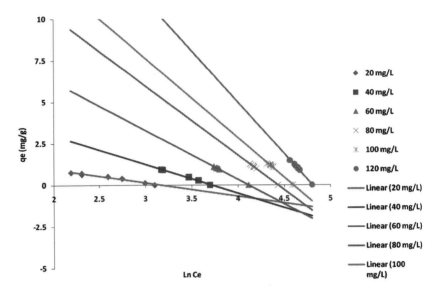

Fig. 34 The linearized Temkin model of RR158 by CAC's. Experimental conditions: room temperature; agitation rate: 200 rpm; biomass dosage: 17.5 g/L; pH: 2

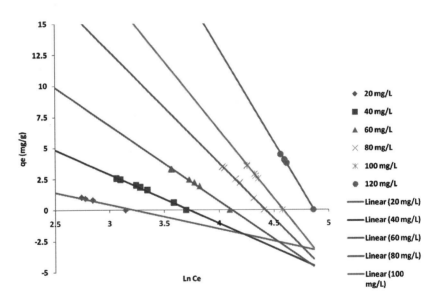

Fig. 35 The linearized Temkin model of RR158 by CAC's. Experimental conditions: room temperature; agitation rate: 200 rpm; biomass dosage: 20 g/L; pH: 2

Table 9 Summary of results of Temkin constant

Initial dye concentration (mg/L)	R^2	b_T	A_T	R^2	b_T	A_T
	Adsorbent dosage: 7.5 g/L			Adsorbent dosage: 15 g/L		
20	0.998	−917.959	0.045	0.997	−1949.309	0.045
40	0.999	−480.243	0.024	0.996	−1124.125	0.024
60	0.999	−352.629	0.017	0.998	−748.737	0.016
80	0.999	−256.106	0.013	0.999	−501.939	0.012
100	1.000	−194.931	0.010	0.998	−427.462	0.010
120	0.999	−168.428	0.008	0.998	−348.708	0.008
	Adsorbent dosage: 10 g/L			Adsorbent dosage: 17.5 g/L		
20	0.999	−1394.244	0.050	0.980	−3093.099	0.042
40	0.999	−662.630	0.025	0.996	−1432.123	0.024
60	0.999	−436.346	0.016	0.999	−840.425	0.016
80	0.999	−329.027	0.012	0.999	−594.142	0.012
100	0.999	−283.897	0.010	0.983	−517.779	0.010
120	0.999	−222.005	0.008	0.998	−398.708	0.008
	Adsorbent dosage: 12.5 g/L			Adsorbent dosage: 20 g/L		
20	0.997	−1637.523	0.045	0.980	−1266.005	0.040
40	–	–	–	0.992	−625.334	0.024
60	0.998	−624.703	0.016	0.992	−406.093	0.016
80	0.999	−396.920	0.012	0.997	−275.194	0.012
100	0.997	−344.682	0.010	0.998	−223.810	0.010
120	0.999	−280.618	0.008	0.999	−164.623	0.008

adsorbent dosage increases such that for adsorbent dosage of 7.5 g/L, the range was 749, and 1172, 1357, 1601 and 2695 for adsorbent dosage of 10.0, 12.5, 15.0, 17.5 and 20 g/L respectively. The A_T values were constant at constant initial dye concentrations with varying adsorbent dosages. The dye solution of initial concentration of 80 mg/L at varying dosages of adsorbent from 7.5 to 20 g/L gives an A_T value equal to 0.12. A_T values were 0.050 for dye concentration of 20 mg/L, 0.024 for dye concentration of 40 mg/L, 0.016–0.017 for dye solution of 60 mg/L, 0.010 for dye solutions of 100 mg/L and 0.008 for dye concentration of 120 mg/L at varying adsorbent dosages of 7.5 to 20 g/L. A_T values increased as the dyes concentration ascended.

4.9 Kinetic Modelling

4.9.1 Pseudo First Order

The adsorption kinetics was investigated for better understanding of the dynamics of adsorption of RR158 onto CAC's and obtaining predictive models that allow estimations of the amount of dye adsorbed with the treatment time. The linearised forms of the models are illustrated in Figs. 36, 37, 38, 39, 40 and 41. A summary of the results for the pseudo first order parameters with the correlation coefficients

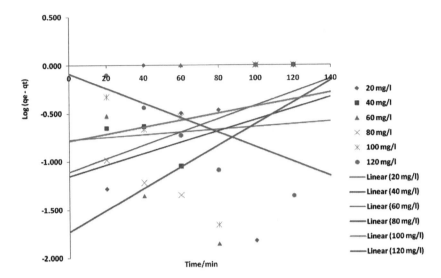

Fig. 36 The linearized pseudo first order kinetics of RR158 by CAC's. Experimental conditions: room temperature; agitation rate: 200 rpm; biomass dosage: 7.5 g/L; pH: 2

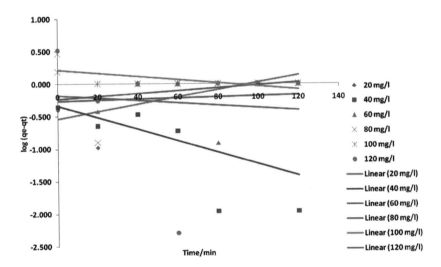

Fig. 37 The linearized pseudo first order kinetics of RR158 by CAC's. Experimental conditions: room temperature; agitation rate: 200 rpm; biomass dosage: 10 g/L; pH: 2

is tabulated in Table 10. The correlation coefficients, R^2 for the first-order kinetic model obtained at the studied concentrations were in the range of 0.002 and 0.416. These low values indicated that the experimental data were not well fitted to the aforementioned model. This may be due to the stirring speed used in the present work (200 rpm) which reduced the film boundary layer [2]. It was also observed

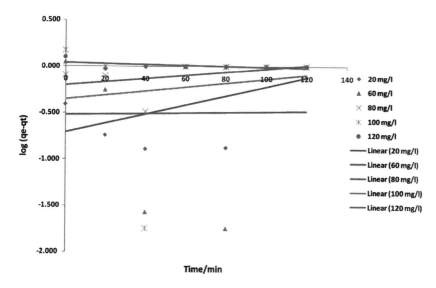

Fig. 38 The linearized pseudo first order kinetics of RR158 by CAC's. Experimental conditions: room temperature; agitation rate: 200 rpm; biomass dosage: 12.5 g/L; pH: 2

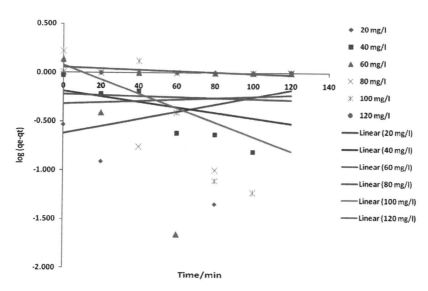

Fig. 39 The linearized pseudo first order kinetics of RR158 by CAC's. Experimental conditions: room temperature; agitation rate: 200 rpm; biomass dosage: 15 g/L; pH: 2

that q_t values computed from this model deviated considerably from the experimental q_t values which indicated that pseudo-first order equation might not be sufficient to describe the mechanism of RR158 and CAC's interactions [11]. The

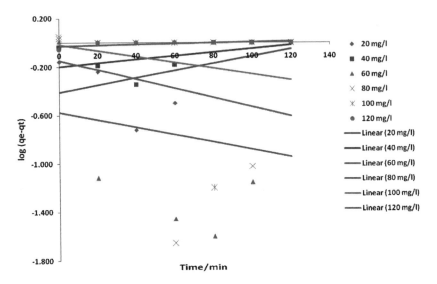

Fig. 40 The linearized pseudo first order kinetics of RR158 by CAC's. Experimental conditions: room temperature; agitation rate: 200 rpm; biomass dosage: 17.5 g/L; pH: 2

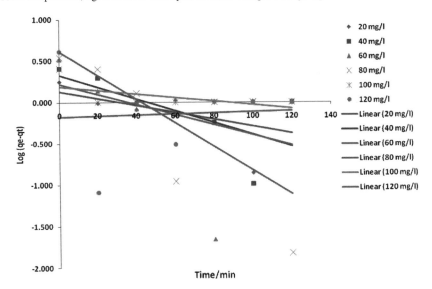

Fig. 41 The linearized pseudo first order kinetics of RR158 by CAC's. Experimental conditions: room temperature; agitation rate: 200 rpm; biomass dosage: 20 g/L; pH: 2

calculated q_t values were lower compared to the experimental q_t values at the various initial dye concentrations and adsorbent dosages. Also, as initial dye concentration increased at a constant dosage of adsorbent, the theoretical q_t values increased such as for adsorbent dosage of 15 g/L, theoretical q_t values increased from 0.2399 to 1.1885 mg/g for increase in dye concentration from 20 to 120 mg/L.

Table 10 Summary of values of rate constant and theoretical q_t values of pseudo first order model

Dye concentration (mg/L)	R^2 value	K_1 (min^{-1})	Theoretical q_t (mg/g)	Experimental q_t (mg/g)	R^2 value	K_1 (min^{-1})	Theoretical q_t (mg/g)	Experimental q_t (mg/g)
	Adsorbent dosage: 7.5 g/L				Adsorbent dosage: 15 g/L			
20	0.005	−0.0023	0.1675	0.348	0.084	−0.0069	0.2399	0.293
40	0.086	−0.0115	0.0703	0.668	0.134	0.0046	0.6457	0.937
60	0.104	−0.0138	0.0778	1.881	0.002	0.0690	1.3772	1.378
80	0.264	0.0253	0.0188	1.941	0.002	0.0290	1.0186	1.019
100	0.048	−0.0069	0.1637	2.296	0.288	0.0161	1.1885	1.652
120	0.274	0.0161	0.0811	4.407	–	–	–	1.307
	Adsorbent dosage: 10 g/L				Adsorbent dosage: 17.5 g/L			
20	0.423	0.0115	0.2904	0.394	0.210	−0.0069	0.9098	0.698
40	0.246	0.0184	0.4592	0.444	0.234	−0.0023	0.6339	0.911
60	0.086	0.0211	0.6669	0.667	0.032	0.0069	0.2642	1.022
80	0.071	−0.0046	0.5848	1.533	0.058	0.0069	0.7096	1.076
100	0.375	0.0046	1.6368	2.894	0.052	0.0046	0.9594	1.130
120	0.007	0.0023	0.6592	3.311	0.021	−0.0069	0.3882	0.876
	Adsorbent dosage: 12.5 g/L				Adsorbent dosage: 20 g/L			
20	0.248	−0.0092	0.1963	0.391	0.272	0.0069	1.3489	1.770
40	0.296	0.0161	1.1534	1.422	0.461	0.0046	1.3489	2.533
60	0.215	0.0516	1.1194	1.120	0.143	0.0115	1.6406	3.289
80	0.177	−0.0023	0.6295	0.809	0.535	0.0069	4.0551	3.459
100	0.019	−0.0046	0.4426	1.498	0.375	0.0184	1.5453	2.556
120	0.228	0.0147	1.2677	1.267	0.256	0.0345	4.0551	4.052

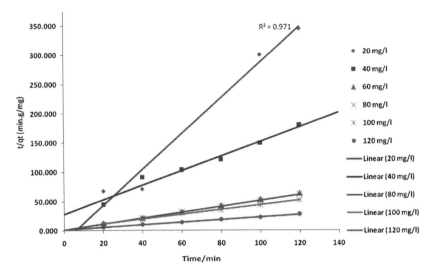

Fig. 42 The linearized pseudo second order kinetics of RR158 by CAC's. Experimental conditions: room temperature; agitation rate: 200 rpm; biomass dosage: 7.5 g/L; pH: 2

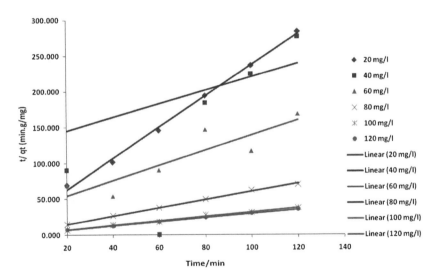

Fig. 43 The linearized pseudo second order kinetics of RR158 by CAC's. Experimental conditions: room temperature; agitation rate: 200 rpm; biomass dosage: 10 g/L; pH: 2

It was also noticed that at adsorbent dosage of 17.5 g/L, the calculated q_t values decreased as dye concentration increased which might be due to the use of insufficiently pretreated coconut shells.

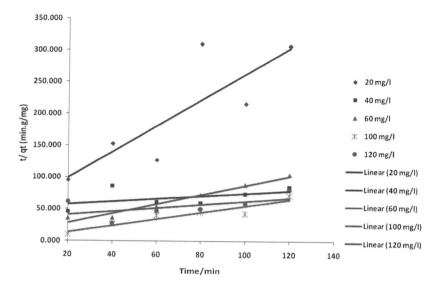

Fig. 44 The linearized pseudo second order kinetics of RR158 by CAC's. Experimental conditions: room temperature; agitation rate: 200 rpm; biomass dosage: 12.5 g/L; pH: 2

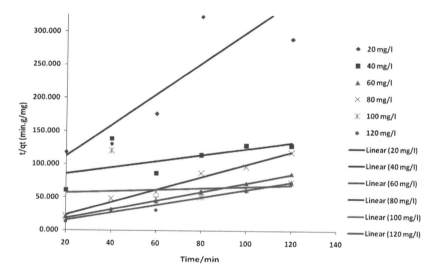

Fig. 45 The linearized pseudo second order kinetics of RR158 by CAC's. Experimental conditions: room temperature; agitation rate: 200 rpm; biomass dosage: 15 g/L; pH: 2

4.9.2 Pseudo Second Order

The linearised form of pseudo second order is shown in Figs. 42, 43, 44, 45, 46 and 47. The computed results of K_2, h and q_t obtained from the second-order

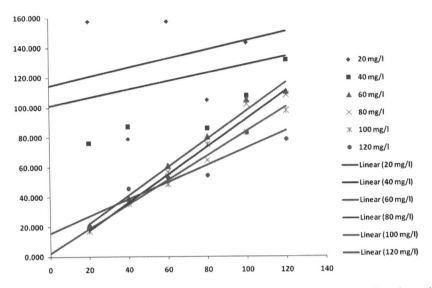

Fig. 46 The linearized pseudo second order kinetics of RR158 by CAC's. Experimental conditions: room temperature; agitation rate: 200 rpm; biomass dosage: 17.5 g/L; pH: 2

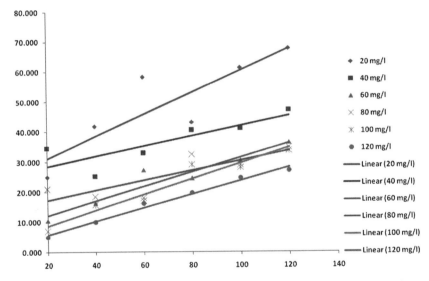

Fig. 47 The linearized pseudo second order kinetics of RR158 by CAC's. Experimental conditions: room temperature; agitation rate: 200 rpm; biomass dosage: 20 g/L; pH: 2

equation are listed in Table 11. The correlation coefficients for the second-order kinetic equation were higher than 0.85 for the batch experiments carried out which indicated that pseudo second order fitted more aptly to pseudo first order. The calculated q_t values predicted from the model agreed very well with the experimental data for example at adsorbent dosage of 7.5 and 20 g/L but with a

Table 11 Summary of results for pseudo second order model

Dye concentration (mg/L)	R^2 values	Theoretical q_c (mg/g)	Experimental q_e (mg/g)	h	K_2 (g/mg min)
		Adsorbent dosage: 7.5 g/L			
20	0.971	0.3279	0.348	−0.0569	−0.5295
40	0.972	0.8078	0.668	0.0352	0.0540
60	0.999	1.9417	1.881	0.6321	0.1677
80	0.999	1.9724	1.941	1.1086	0.2850
100	0.995	2.4570	2.296	0.2990	0.0495
120	0.999	4.5872	4.407	0.8881	0.0422
		Adsorbent dosage: 10 g/L			
20	0.997	0.4575	0.394	0.0525	0.2511
40	0.069	1.6313	0.444	0.0053	0.0020
60	0.785	0.9407	0.667	0.0301	0.0341
80	0.997	1.7452	1.533	0.3007	0.0987
100	0.997	3.2362	2.894	1.1074	0.1057
120	0.999	3.4965	3.311	0.9066	0.0742
		Adsorbent dosage: 12.5 g/L			
20	0.692	0.4895	0.391	0.0171	0.0713
40	0.226	4.6083	1.422	0.0187	0.0009
60	0.969	1.3477	1.120	0.0750	0.0413
80	0.954	0.1898	0.809	0.0016	0.0457
100	0.880	1.9646	1.498	0.2642	0.0684
120	0.336	3.7037	1.267	0.0281	0.0020
		Adsorbent dosage: 15 g/L			
20	0.766	0.4277	0.293	0.0152	0.0831
40	0.942	2.1008	0.937	0.0132	0.0030
60	0.996	1.4925	1.378	0.2196	0.0986
80	0.984	1.0482	1.019	0.2402	0.2186
100	0.018	8.3333	1.652	0.0184	0.0003
120	0.996	1.7271	1.307	0.2846	0.0954
		Adsorbent dosage: 17.5 g/L			
20	0.099	3.2895	0.698	0.0087	0.0008
40	0.030	3.6364	0.911	0.0099	0.0007
60	0.984	1.0549	1.022	0.3246	0.2917
80	0.970	1.0661	1.076	−0.7289	−0.6413
100	0.985	1.2210	1.130	0.4751	0.3187
120	0.891	1.7422	0.876	0.0641	0.0211
		Adsorbent dosage: 20 g/L			
20	0.757	2.7248	1.770	0.0420	0.0057
40	0.680	5.8824	2.533	0.0399	0.0012
60	0.913	4.1322	3.289	0.1378	0.0081
80	0.693	6.0606	3.459	0.0715	0.0019
100	0.930	3.8168	2.556	0.2938	0.0202
120	0.986	4.4444	4.052	0.7987	0.0404

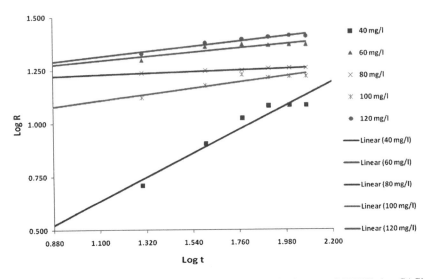

Fig. 48 The linearized intra particle diffusion adsorption isotherms of RR158 by CAC's. Experimental conditions: room temperature; agitation rate: 200 rpm; biomass dosage: 7.5 g/L; pH: 2

few exceptions certain dosages of adsorbent. These indicated that the adsorption system studied was very consistent with the second-order kinetic model. The second-order rate constant K_2 was in the range of 0.0003–0.2917 g/(mg) (min) at the given initial dye concentrations while the predicted values of h were in the range of 0.0053–1.1086 mg/g min for the batch experiments carried out.

4.9.3 Intra Particle Diffusion Model

The linearised intra-particle diffusion model deduced for this work are illustrated in Figs. 48, 49, 50, 51, 52 and 53 followed by a summary of results giving the aforesaid parameters in Table 12. The R^2 values were higher than 0.800 for most of the batch experiments such as for dye concentrations of 40, 80, 100 and 120 mg/L when adsorbent dosage was 7.5 g/L, and the R^2 values were 0.827, 0.906, 0.808 and 0.827 for dye concentrations of 20, 40, 60, 80 mg/L respectively when adsorbent dosage was 20 g/L. These high values of correlations indicated that the experimental data was well fitted to the model. However, the R^2 values were in the range of 0.040–0.799 for a few batch experiments. The k_{id} values were in the range of 0.029–37.325 h^{-1} whereby high values of k_{id} demonstrated that the dyes molecules were transported to the external surface of the CAC's through film diffusion and its rate was fast while the low values indicated that rate of transfer of dye molecules from the solution to the adsorbent was low. The low values of k_{id} might be due to the use of improperly chemically prepared adsorbent.

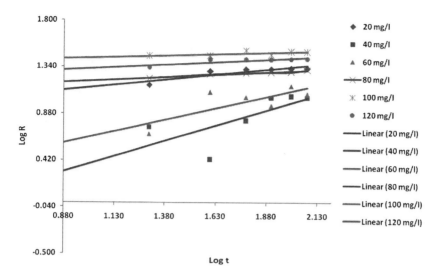

Fig. 49 The linearized intra particle diffusion adsorption isotherms of RR158 by CAC's. Experimental conditions: room temperature; agitation rate: 200 rpm; biomass dosage: 10 g/L; pH: 2

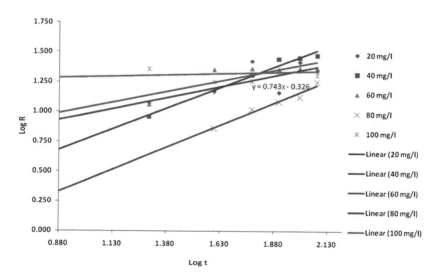

Fig. 50 he linearized intra particle diffusion adsorption isotherms of RR158 by CAC's. Experimental conditions: room temperature; agitation rate: 200 rpm; biomass dosage: 12.5 g/L; pH: 2

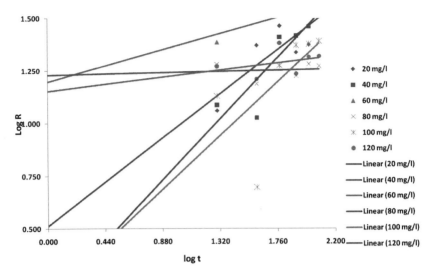

Fig. 51 The linearized intra particle diffusion adsorption isotherms of RR158 by CAC's. Experimental conditions: room temperature; agitation rate: 200 rpm; biomass concentration: 15 g/L; pH: 2

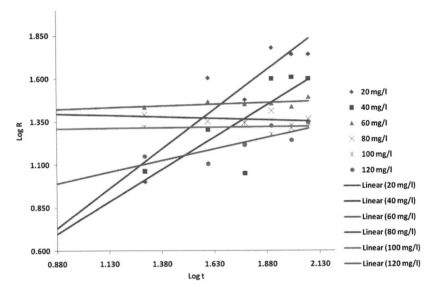

Fig. 52 The linearized intra particle diffusion adsorption isotherms of RR158 by CAC's. Experimental conditions: room temperature; agitation rate: 200 rpm; biomass dosage: 17.5 g/L; pH: 2

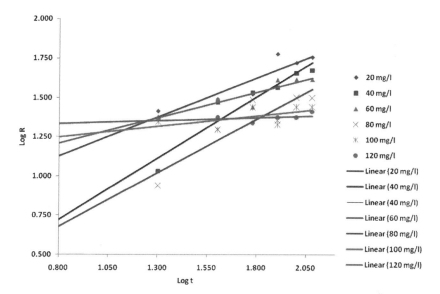

Fig. 53 The linearized intra particle diffusion adsorption isotherms of RR158 by CAC's. Experimental conditions: room temperature; agitation rate: 200 rpm; biomass dosage: 20 g/L; pH: 2

4.10 Batch Column Studies

4.10.1 Effect of Bed Height

The observed curves of RR158 adsorption on CAC's at three different bed heights are displayed in Fig. 54, plotted as the percentage dye removal versus time and Fig. 55 shows the breakthrough curves. The uptake of RR158 increased from 73.60% to 92.49% with increase in bed height from 6 to 10 cm. This rise in dye uptake was due to the increase in mass of adsorbent in the column resulting in an increase in the surface area of adsorbent, which provided more binding sites for adsorption [1].

Results indicated that volume varies with bed depth. This displacement of the front of adsorption with the increase in depth can be explained by mass transfer phenomena that take place in this process. When bed depth is reduced, axial dispersion phenomena predominates in the mass transfer and reduces the diffusion of species. The solute thus does not have enough time to diffuse into the whole of the adsorbent mass thus leading to a reduction in dye uptake.

The breakthrough curves (Fig. 55) had high R^2 values of 0.972 for bed height of 6 and 8 cm and 0.990 for bed height of 10 cm, and so followed the characteristics "S" shape profile produced in an ideal adsorption system. Less sharp breakthrough curves were obtained at higher mass of adsorbent. The variation in breakthrough

Table 12 Summary of results of intra particle diffusion model

Dye concentration (mg/L)	R^2 value of graph	a	k_{id} (h^{-1})
	Adsorbent dosage: 7.5 g/L		
20	0.040	−0.528	37.325
40	0.942	0.508	1.052
60	0.751	0.088	12.853
80	0.970	0.032	15.596
100	0.829	0.130	9.226
120	0.939	0.108	15.704
	Adsorbent dosage: 10 g/L		
20	0.799	0.198	8.620
40	0.488	0.598	1.641
60	0.586	0.449	1.563
80	0.947	0.093	12.794
100	0.397	0.053	23.659
120	0.740	0.099	16.634
	Adsorbent dosage: 12.5 g/L		
20	0.510	0.370	4.020
40	0.357	1.194	0.096
60	0.703	0.357	4.710
80	0.801	0.743	0.472
100	0.031	0.046	17.458
120	0.107	0.739	5.248
	Adsorbent dosage: 15 g/L		
20	0.694	0.474	3.266
40	0.795	0.657	1.426
60	0.704	0.173	15.776
80	0.010	0.013	16.943
100	0.384	0.578	1.510
120	0.120	0.076	14.191
	Adsorbent dosage: 17.5 g/L		
20	0.807	0.920	0.832
40	0.633	0.751	1.084
60	0.295	0.037	24.434
80	0.082	−0.035	26.607
100	0.016	0.010	19.861
120	0.645	0.266	5.623
	Adsorbent dosage: 20 g/L		
20	0.827	0.496	5.383
40	0.906	0.781	1.253
60	0.808	0.323	8.892
80	0.827	0.681	1.358
100	0.354	0.134	0.057
120	0.220	0.038	19.999

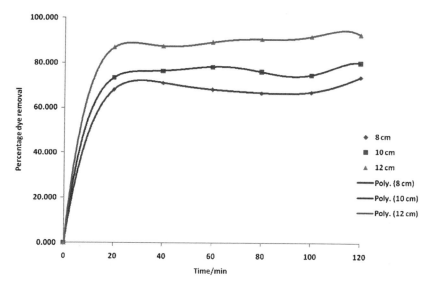

Fig. 54 Graph of dye removal versus time for batch column experiment. Experimental conditions: pH: 2, volume of dye: 250 mL, initial dye concentration of dye: 120 mg/L

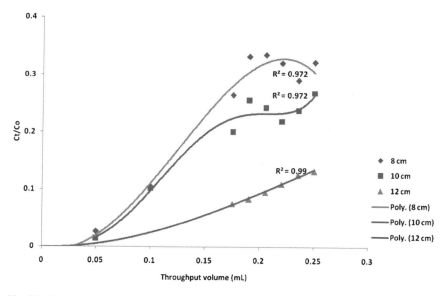

Fig. 55 Breakthrough curves of adsorption of RR158 dye on CAC's at various bed heights. Experimental conditions: pH: 2, input volume of dye: 250 mL, initial dye concentration: 120 mg/L

shape with CAC's mass is mainly due to the relatively large adsorption zone that is CAC's near the bottom of the adsorption column, which comes into contact with dye solution after CAC's near the top of column, is exhausted. Therefore, the mass transfer zone (MTZ) has its own characteristics that are dependent upon the nature

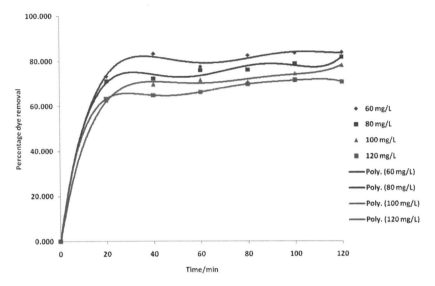

Fig. 56 Dye removal at varying initial dye concentration. Experimental conditions: pH: 2, volume of dye input: 150 mL, bed height: 6 cm

of adsorbent–adsorbate interactions in addition to the experimental conditions such as mass of adsorbent and initial dye concentration.

4.10.2 Column Studies: Effect of Initial Dye Concentration

The effect of a variation from 60 to 120 mg/L of the inlet concentration of the solution set at constant volume and bed height is shown in Fig. 56 followed by Fig. 57 illustrating the breakthrough curves. A rise in the inlet dye concentration caused a reduction in the dye removal before the packed bed gets saturated. It was also observed that the time taken to achieve adsorption equilibrium point was independent of the variation in dye concentration. Thus, dye solution irrespective of their concentration had an equilibrium time of 40 min at the given bed heights and adsorbent dosages.

4.10.3 Column Studies: Effect of Input Volume of Dye Solution

Adsorption of RR158 was studied at the input volume of 100, 150, 200 and 250 mL in order to investigate the effect of input volume on the adsorption behaviour. Initial dye concentration, pH and bed height were maintained constant. The experimental results of the effect of input volume are illustrated in Fig. 58 and the breakthrough curves shown in Fig. 59. An increase in dye uptake was noticed with a decrease in input dye solution. The removal was 92.63, 83.28, 82.36 and 60.00% for input volumes of 100, 150, 200 and 250 mL, respectively. The high

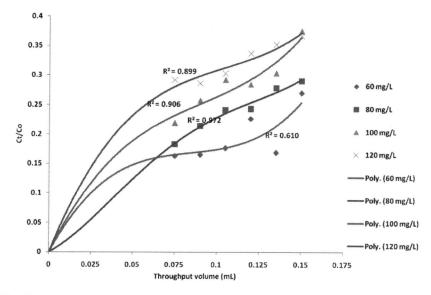

Fig. 57 Breakthrough curve of RR158 dye adsorption on CAC's for various concentrations of dye solution. Experimental conditions: pH: 2, volume of solution: 150 mL, bed height: 6 cm

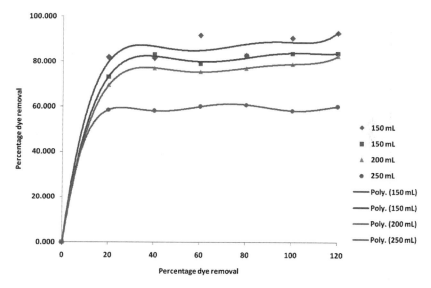

Fig. 58 Effect of input volume of dye solution on dye removal. Experimental condition: pH: 2, bed height: 6 cm, initial dye concentration: 60 mg/L

R^2 values of the breakthrough curves of 0.806, 0.900, 0.823 and 0.958 for input volumes of 100, 150, 200 and 250 mL, respectively, bear evidence of good correlation with the experimental data.

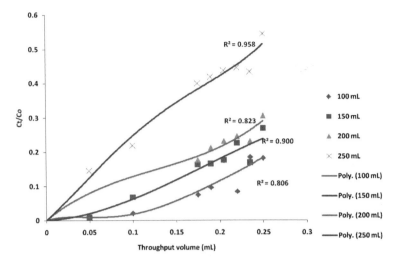

Fig. 59 Breakthrough curves of RR158 adsorption on CAC's for various input volume of solution. Experimental conditions: pH: 2, bed height: 6 cm, initial dye concentration: 60 mg/L

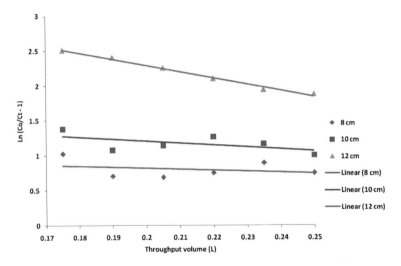

Fig. 60 Linearised Thomas model of RR158 adsorption on CAC's showing effect of bed height. Experimental conditions: volume of dye solution: 250 mL, pH: 2 and concentration of dye solution: 120 mg/L

4.10.4 Thomas Model to Batch Column Experiments

The linearised Thomas models for the data in this work are shown in Figs. 60, 61 and 62 at varying bed heights, varying input volume and varying initial dye

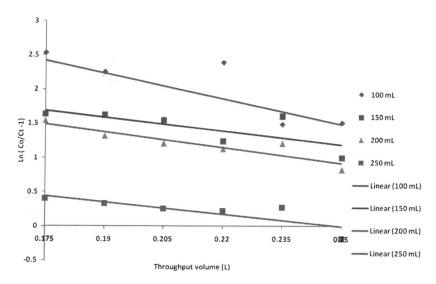

Fig. 61 Linearised Thomas model of RR158 adsorption on CAC's for varying input volume. Experimental conditions: concentration of dye: 60 mg/L, pH: 2 and bed height: 6 cm

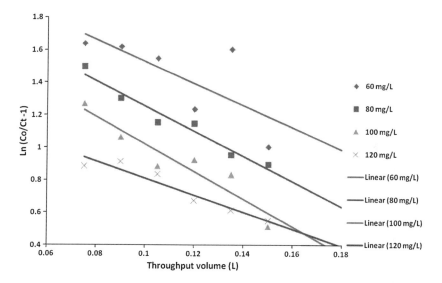

Fig. 62 Linearised Thomas model of RR158 adsorption on CAC's for varying initial dye concentration. Experimental conditions: volume of dye solution: 150 mL, pH: 2 and bed height: 6 cm

concentration respectively. A summary of the results of the Thomas parameters determined from the Thomas model is displayed in Table 13. It was observed that the Thomas constant was directly proportional to both the bed height and the input

Table 13 Summary of results of Thomas model parameters

	R^2 values	k_T (L/min mg) $\times\ 10^{-5}$	q_0 (mg/g)
Present study			
Bed height/cm	Effect of bed height		
8	0.190	2.432	0.446
10	0.329	4.837	0.272
12	0.989	15.435	0.157
Input volume/mL	Effect of input volume		
100	0.518	17.292	0.105
150	0.525	14.083	0.121
200	0.825	16.922	0.105
250	0.669	20.826	0.071
Initial dye concentration (mg/L)	Initial dye concentration (mg/L)		
60	0.525	14.083	0.093
80	0.955	12.073	0.099
100	0.888	10.566	0.105
120	0.917	5.438	0.146
	R^2	k_T (L/min mg) $\times\ 10^{-3}$	q_0 (mg/g)
From literature [1]			
Initial dye concentration/mg/L			
57.0	0.991	5.19	14.12
79.0	0.996	3.30	20.96
100.0	0.989	2.04	24.58
Mass of adsorbent/g			
0.1	0.986	22.71	5.49
0.2	0.993	22.33	7.04
0.3	0.998	13.51	8.77
Flow rate (mL/min)			
6.82	0.998	10.10	10.10
5.19	0.994	7.51	12.08
3.77	0.991	5.19	14.12

volume of dye solution. The k_T values increased from 2.432E−05 L/min mg to 15.435E−05 L/min mg as bed height increased from 6 to 10 cm and from 14.083E−05 to 20.826E−05 L/min mg when the input volume of dye increased from 150 to 250 mL.

However, as far as the initial dye concentration was concerned, the Thomas constant was inversely relative such that the k_T values decreased from 14.083E−05 to 5.438E−05 L/min mg as dye concentration increased from 60 to 120 mg/L. The calculated q_0 values were inversely proportional to the bed heights and input volume but directly relative to initial dye concentration. The q_0 values increased from 0.093 to 0.146 mg/g for rise in dye concentration from 60 to 120 mg/L, and decreased from 0.446 to 0.157 mg/g for rise in bed height from 6 to 10 cm and decreased from 0.105 to 0.071 mg/g for rise in input volume from 100 to 250 mL.

Table 14 Summary of results of Langmuir model and DR model

Dye concentration (mg/L)	SAE Langmuir model	ARE	RSS	SAE DR model	ARE	RSS
Adsorbent dosage: 7.5 g/L						
20	0.346	0.995	0.120	0.190	0.546	0.036
40	0.614	0.919	0.377	0.444	0.664	0.197
60	1.509	0.802	2.277	0.671	0.356	0.450
80	1.606	0.827	2.580	0.440	0.227	0.194
100	1.986	0.865	3.944	1.670	0.727	2.788
120	3.469	0.787	12.033	3.104	0.704	9.632
Adsorbent dosage: 10 g/L						
20	0.334	0.847	0.111	0.362	0.920	0.131
40	0.430	0.969	0.185	0.444	1.000	0.197
60	0.627	0.941	0.394	0.665	0.996	0.442
80	1.249	0.815	1.561	1.366	0.891	1.865
100	1.987	0.686	3.947	1.964	0.678	3.855
120	2.594	0.784	6.730	2.710	0.818	7.344
Adsorbent dosage: 12.5 g/L						
20	0.332	0.848	0.110	0.361	0.923	0.130
40	–	–	–	1.356	0.954	1.839
60	0.797	0.712	0.636	1.057	0.944	1.117
80	0.806	0.996	0.650	0.809	1.000	0.654
100	1.081	0.721	1.168	1.156	0.772	1.336
120	1.213	0.958	1.472	1.264	0.997	1.596
Adsorbent dosage: 15 g/L						
20	0.245	0.836	0.060	0.265	0.904	0.070
40	0.841	0.898	0.708	0.701	0.748	0.491
60	1.042	0.756	1.086	1.036	0.752	1.073
80	0.856	0.840	0.733	0.934	0.917	0.872
100	1.564	0.946	2.445	1.620	0.980	2.623
120	0.993	0.760	0.986	1.094	0.837	1.197
Adsorbent dosage: 17.5 g/L						
20	0.615	0.881	0.378	0.629	0.901	0.395
40	0.800	0.878	0.640	0.824	0.904	0.679
60	0.729	0.713	0.531	0.792	0.775	0.627
80	0.820	0.762	0.673	0.866	0.805	0.750
100	0.895	0.792	0.801	0.966	0.855	0.933
120	0.687	0.784	0.472	0.778	0.889	0.606
Adsorbent dosage: 20 g/L						
20	1.262	0.713	1.594	1.149	0.649	1.320
40	2.257	0.891	5.095	2.279	0.900	5.194
60	2.450	0.745	6.003	2.448	0.744	5.992
80	3.132	0.906	9.812	3.200	0.925	10.242
100	1.828	0.715	3.342	1.940	0.759	3.762
120	3.079	0.760	9.481	3.236	0.799	10.470

Table 15 Summary of results of error analysis of Pseudo first and second order

Dye concentration (mg/L)	SAE	ARE	RSS	SAE	ARE	RSS
	Pseudo first order			Pseudo second order		
Adsorbent dosage: 7.5 g/L						
20	0.181	0.519	0.033	0.020	0.058	0.000
40	0.598	0.895	0.357	0.139	0.209	0.020
60	1.803	0.959	3.252	0.060	0.032	0.004
80	1.922	0.990	3.695	0.031	0.016	0.001
100	2.132	0.929	4.547	0.161	0.070	0.026
120	4.326	0.982	18.713	0.180	0.041	0.032
Adsorbent dosage: 10 g/L						
20	0.104	0.263	0.011	0.064	0.161	0.004
40	0.015	0.034	0.000	1.187	2.674	1.410
60	0.000	0.000	0.000	0.274	0.410	0.075
80	0.948	0.619	0.899	0.212	0.138	0.045
100	1.257	0.434	1.581	0.342	0.118	0.117
120	2.652	0.801	7.032	0.186	0.056	0.034
Adsorbent dosage: 12.5 g/L						
20	0.195	0.498	0.038	0.099	0.252	0.010
40	0.269	0.189	0.072	3.186	2.241	10.153
60	0.001	0.001	0.000	0.228	0.203	0.052
80	0.180	0.222	0.032	0.619	0.765	0.383
100	1.055	0.705	1.114	0.467	0.312	0.218
120	0.001	0.001	0.000	2.437	1.923	5.938
Adsorbent dosage: 15 g/L						
20	0.053	0.181	0.003	0.135	0.460	0.018
40	0.291	0.311	0.085	1.164	1.242	1.354
60	0.001	0.001	0.000	0.115	0.083	0.013
80	0.000	0.000	0.000	0.029	0.029	0.001
100	0.464	0.281	0.215	6.681	4.044	44.640
120	1.307	1.000	1.708	0.420	0.321	0.176
Adsorbent dosage: 17.5 g/L						
20	0.212	0.303	0.045	2.592	3.713	6.716
40	0.277	0.304	0.077	2.725	2.992	7.428
60	0.758	0.741	0.574	0.033	0.032	0.001
80	0.366	0.341	0.134	0.010	0.0010	0.000
100	0.171	0.151	0.029	0.091	0.081	0.008
120	0.488	0.557	0.238	0.866	0.989	0.750
Adsorbent dosage: 20 g/L						
20	0.421	0.238	0.177	0.955	0.539	0.912
40	1.184	0.467	1.402	3.349	1.322	11.218
60	1.648	0.501	2.717	0.843	0.257	0.711
80	0.596	0.172	0.355	2.602	0.752	6.768
100	1.011	0.395	1.022	1.261	0.493	1.590
120	0.003	0.001	0.000	0.392	0.097	0.154

Table 16 Summary of results of error analysis of Thomas model

	SAE	ARE	RRS
Effect of bed height (cm)			
6	0.348	3.551	0.121
8	0.192	2.400	0.037
10	0.080	1.039	0.006
Effect of input volume (mL)			
100	0.081	3.375	0.007
150	0.083	2.184	0.007
200	0.059	1.283	0.003
250	0.029	0.690	0.001
Initial dye concentration (mg/L)			
60	0.055	1.447	0.003
80	0.058	1.415	0.003
100	0.055	1.100	0.003
120	0.088	1.517	0.008

4.11 Data Analysis

A summary of the results of error analysis carried out for the equilibrium and
kinetics models of adsorption process are given Tables 14, 15 and 16. The values
of RSS, SAE and ARE about Langmuir model at most of the constant dosages of
adsorbent used were smallest between the two isotherms of Langmuir and DR.
It could be concluded that the Langmuir model was the best fit to the experimental
data. For the kinetics model, it was observed that at low dye concentrations and
high dosages of adsorbent, the errors of pseudo first order were smaller compared
to the second order. However at low adsorbent dosages of 7.5 and 10 g/L, the error
values of pseudo second order were lower indicating that it was the best fit to
experimental data.

The RSS, SAE and ARE values from Thomas model showed that as bed height
and input volume increased, the errors decreased while for increase in dye concentration, there was an increase in the error values.

5 Conclusions

The past 10 years have seen a developing interest in the preparation of low-cost
adsorbents as alternatives to activated carbons in water and wastewater treatment
processes. This work has studied the effect of chemically pretreated *C. nucifera* L.
shells in the removal of RR158 dye from synthetic solutions in batch experiments
and fixed bed column under a variety of operating conditions. The adsorbent
preparation experiment was 89.3% efficient with little loss of coconut shells. The
RR158 removal efficiency of the adsorbent increased with increase in time,

adsorbate concentration, and decreased with increase in pH (maximal RR158 removal at pH 2) and adsorbent dose. It is inferred that the adsorption of RR158 onto CAC's in a fixed bed column is more effective than the batch system as the removal efficiencies were in the range of 80–90% for the batch column and around 50% for the batch experiments. Also, the equilibrium time was around 40 min at high dye concentrations of 80, 100 and 120 mg/L but low dye concentrations of 20 and 40 mg/L reached equilibrium point in 120 min.

The experimental equilibrium data values from the batch experiments fitted well to the Langmuir, Freundlich, Dubinin–Radushkevich and Temkin adsorption isotherm models which were supported by their high correlation coefficients, all above 0.850. The constants of the isotherms were determined and these values are most useful to design single-stage batch absorbers for the removal of RR158 dye using CAC's. The adsorption kinetics agreed more consistently to pseudo second order model than the pseudo first order which was supported by the R^2 values and the error analysis carried out for both kinetic models. It has been observed that intra-particle diffusion also played an important part in the adsorption mechanism thus contributing to higher uptake of dye molecules from the synthetic solution. The adsorption of RR158 onto CAC's using the columns exhibited a characteristic "S" shape and was effectively represented by the Thomas model. The adsorption capacity of the adsorbent of *C. nucifera* L. increased as the initial dye concentration increased due to the concentration gradient. The use of high flow rates reduced the time contact between RR158 in the solution with the CAC's, thus allowing less time for adsorption to occur, leading to an early breakthrough of RR158.

It may be concluded that dye uptake and removal from textile effluents using the adsorption technique can be beneficial in terms of high adsorption efficiency and also in terms of production of less wastes as the CAC's may be potentially recovered for reuse up to a certain number of cycles. The adsorption process of dyes with pretreated *C. nucifera* L. shells can be hence considered to be a sustainable option for remediating dye-laden effluents. Further work needs now be geared towards upscaling the present findings and results into designing pilot-scale adsorption units.

References

1. Al-Ghouti MA, Khraisheh MAM, Ahmad MN et al (2006) Microcolumn studies of dye adsorption onto manganese oxides modified diatomite. J Hazard Mater 16:45–65
2. Babu BV, Gupta S (2008) Adsorption of Cr(VI) using activated neem leaves: kinetic studies. Adsorption 14:85–92
3. Dang VBH, Doan HD, Dang-Vu T et al (2009) Equilibrium and kinetics of biosorption of cadmium(II) and copper(II) ions by wheat straw. Bioresour Technol 100:211–219
4. Demirbas E, Kobya M, Senturk E et al (2004) Adsorption kinetics for the removal of chromium (VI) from aqueous solutions on the activated carbons prepared from agricultural wastes. Water SA 30:52–60
5. Igwe JC, Abia AA (2006) A bioseparation process for removing heavy metals from waste water using biosorbents. Afr J Biotechnol 5:1167–1179

6. Kansal SK, Singh M, Sud D (2006) Studies on photodegradation of two commercial dyes in aqueous phase using different photocatalysts. Photodegradation 16:117–127
7. Khehra MS, Saini HS, Sharma DK et al (2006) Biodegradation of azo dye C.I. Acid Red 88 by an anoxiceaerobic sequential bioreactor. Dyes Pigments 70:1–7
8. Kidwai M, Mohan R (2005) Green chemistry: an innovative technology. Found Chem 7:269–287
9. Loukidou MX, Zouboulis AI, Karapantsios TD et al (2004) Equilibrium and kinetic modeling of chromium (VI) biosorption by *Aeromonas caviae*. Colloids Surf A Physicochem Eng Aspects 242:93–104
10. Molen J (2008) Treatability studies on the wastewater from a dye manufacturing industry. Minerals Eng 24:48–68
11. Pino GHN, Mesquita LMS, Torem ML et al (2006) Biosorption of cadmium by green coconut shell powder. Minerals Eng 19:380–387
12. Rodrigues LA, Maschio LJ, Silva RED et al (2010) Adsorption of Cr(VI) from aqueous solution by hydrous zirconium oxide. J Hazard Mater 173:630–636
13. Sarin V, Pant KK (2006) Removal of chromium from industrial waste by using eucalyptus bark. Bioresour Technol 97:15–20
14. Semerjian L (2010) Equilibrium and kinetics of cadmium adsorption from aqueous solutions using untreated *Pinus halepensis* sawdust. J Hazard Mater 173:236–242
15. Shu Y, Li L, Zhang Q et al (2010) Equilibrium, kinetics and thermodynamic studies for sorption of chlorobenzenes on CTMAB modified bentonite and kaolinite. J Hazard Mater 173:47–53
16. Sud D, Mahajan G, Kaur MP (2008) Agricultural waste material as potential adsorbent for sequestering heavy metal ions from aqueous solutions—a review. Bioresour Technol 99:6017–6027
17. Terdkiatburana T, Wang S, Tadé MO (2009) Adsorption of heavy metal ions by natural and synthesised zeolites for wastewater treatment. Int J Environ Waste Manag 3:1126–1148
18. Vijayaraghavan K, Yun YS (2008) Bacterial biosorbents and biosorption. J Biotechnol Adv 26:266–291